THE GREAT REVERSAL

T0221193

How We Let Technology Take Control of the Planet

Every day, we are presented with new technologies that can influence human thought and action, such as psychopharmaceuticals, new generation performance-enhancing drugs, elective biotechnology, and gastric bypass surgery. Have we let technology go too far in this respect? In *The Great Reversal*, David Edward Tabachnick contends that this question may not be unique to contemporary society. Through an assessment of the great works of philosophy and politics, Tabachnick explores the largely unrecognized history of technology as an idea.

The Great Reversal takes the reader back to Aristotle's ancient warning that humanity should never allow technical thinking to cloud our judgment about what makes for a good life. It then charts the path of how we began to relinquish our deeply rooted intellectual and practical capacities that used to allow us to understand and regulate the role of technologies in our lives. As the rise of technology threatens our very humanity, Tabachnick emphasizes that we still may have time to recover and develop these capacities – but we must first decide how far we want to allow technology to determine our existence and our future.

DAVID EDWARD TABACHNICK is an associate professor in the Department of Political Science at Nipissing University.

The Great Reversal

How We Let Technology Take Control of the Planet

DAVID EDWARD TABACHNICK

UNIVERSITY OF TORONTO PRESS
Toronto Buffalo London

ISBN 978-0-8020-9263-2 (cloth)
ISBN 978-0-8020-9469-8 (paper)

Library and Archives Canada Cataloguing in Publication

Tabachnick, David
The great reversal : how we let technology take control of the planet / David
Edward Tabachnick.

Includes bibliographical references and index.
ISBN 978-0-8020-9263-2 (bound). – ISBN 978-0-8020-9469-8 (pbk.)

1. Technology and civilization.   2. Technology – Philosophy.   3. Technology
– Social aspects.   4. Technology – Moral and ethical aspects.   5. Virtue.

CB478.T22 2013   303.48'3   C2012-906125-5

University of Toronto Press acknowledges the financial assistance to its
publishing program of the Canada Council for the Arts and the Ontario
Arts Council.

University of Toronto Press acknowledges the financial support of the
Government of Canada through the Canada Book Fund for its publishing
activities.

Contents

Foreword

A few years ago, the great contemporary philosopher of technology Andrew Feenberg wrote: "We are well aware that we are a technological society, and not just because we use so many devices but also in our spirit and our way of life."[1] To be honest, I have never been quite sure to whom this sentence refers, and by this I do not mean simply that there is no such thing as "we" anymore. Even if we (there it is again) accept that "we" means those of us who live in something like the modern way, I am not sure that we generally are "well aware" of what I think is otherwise the basic truth of Feenberg's claim. To be sure, few will have failed to notice that our everyday practices are completely mediated by devices of one sort or another. Some would suggest that such mediation is original and necessary to human experience in any case, and that our awareness of this condition has simply been heightened by the apparent novelty of various digital prostheses. These have prompted anxious popular discussion of the "impact" of digital technology on our "quality of life" in various domains (work, commerce, courtship, parenting, and childhood), but I am not sure this approaches what is contemplated by the suggestion that technology inheres in "our spirit and our way of life." If pressed, I suspect most of us would respond to the question "What sort of society do you live in?" with the answer "a democracy." Of course, here is the rub: most of us so closely identify democracy with technology that we have become unable to distinguish

1 Andrew Feenberg, *Heidegger and Marcuse: The Catastrophe and Redemption of History* (New York: Routledge, 2005), 6.

between them, or to define either one of them in terms that do not presumptively invoke the other.

The book you now hold in your hands, or see on your screen, aims to correct that, by refusing to collapse the distinction between technology and politics. It seeks to provoke a heightened, critical awareness of just how deep our collective relationship with technology goes, and what is at stake in it. In so doing, it speaks to the cruel irony at the centre of the technological condition: that in our efforts to exercise control over the contingencies that define our existence in the world, we have relinquished our capacity to exert control over the place of technology in our public and private lives. It is not just a matter of the joke that email plays on us ("Here is a great new tool for interpersonal communication that will make you more productive, better organized, and give you tons of free time you never had before!"), but rather that we cannot imagine a way out of our current economic and environmental predicaments that does not involve massive investment in precisely the same technological enterprise that produced the situation in the first place. This, we might say, is the broken heart at the core of David Tabachnick's prescient and insightful book, a broken heart he seeks to mend.

In a radical departure from present orthodoxies in social studies of technology, Tabachnick does not treat technology as a mere ensemble of devices whose meaning derives exclusively from the specific contexts in which their design, application, and use patterns unfold. In the pages that follow, technology is instead presented as a particular way of being in the world that has eclipsed other ways of being in the world. In particular, technology is the prevailing and definitive way people in the frantic countries of industrial, post-industrial, and informational capitalism have come to reckon with the uncertainties the world necessarily presents to them. Tabachnick's concern is for what is overshadowed in this eclipse, and for whether and how it might be brought back into the light. For him, technology is not just a historical or sociological question but, moreover, a deeply philosophical one.

The way of being in the world that has been eclipsed by technology is characterized by routine engagement of the human capacity for practical judgment, a capacity central to the possibility of ethical and political life. The story Tabachnick tells is the history of the contest, declared or otherwise, between *techne* and *phronesis*, between the useful arts and the sort of judgment that might direct these towards truly good purposes. It is a story of reversal, in which the priority properly assigned to political and ethical judgment has somehow given way to the perceived imperatives of technological development, and it is a story about what this turn of events means for our prospects. In this, Tabachnick joins a long line of storytellers who have tried in various ways to narrate the technological predicament – Heidegger, Marcuse, Ellul, Illich, Grant, Winner, Borgmann – thinkers on whose shoulders he stands but beyond whose historical horizons he clearly sees. For, whatever truth there might be to claims that human beings have been technological since they first engaged productively with the world outside themselves, it is hard to deny that the extent and intensity of contemporary technological activity lends a distinctive practical urgency to the question of how we are to live with the technologies we make, and that make us what we are.

As Tabachnick puts it, "Somewhere along the line we began to relinquish the deeply rooted intellectual and practical capacities that allow us to understand and regulate the role of technology in our lives." The primary endeavour here is, accordingly, forensic: when, where, and how did it happen, and who did it? Tabachnick's signature proposition, substantiated by a magnificent tour through the history of Western political thought, is that "our idea about technology and how it has changed is largely responsible for the creation and growth of our technological society." This central motif will undoubtedly provoke as many readers as it convinces, for the claim that our idea about a thing – especially as rendered in philosophical texts – determines the reality of that thing in the world fairly cries out for critical confrontation. This goes double for something as irreducibly material as technology. Nevertheless, this is Tabachnick's bold gambit, and the careful manner in which he pursues it should be enough to make serious readers reflect upon why they think what they do about the technology that surrounds them. Tabachnick's sober, sobering, and detailed attention to the material challenges of contemporary technology confirm that he does not imagine that we can do away with technology simply by doing away with the thought of it, as materialist critics might otherwise be inclined to charge. Instead, like all autopsies, his forensic treatment of the career of technology in Western political thought might be exactly what we need to get past the tragedy that put the body on the table, and to move on in light of what we have learned about what has brought us here.

DARIN BARNEY
Canada Research Chair in Technology & Citizenship
McGill University

Acknowledgments

The first concrete plan for this book was conceived during my year as a Fulbright Visiting Research Chair. For this, I gratefully acknowledge the support of the people at the Fulbright Foundation and at Fulbright Canada. A draft was written during a sabbatical year from Nipissing University.

My ideas for this project came over many years, but began to take form under the supervision and guidance of Tom Darby and Randy Newell. I give special thanks to my close colleague and friend Toivo Koivukoski for conversations that helped me sharpen and clarify my arguments. Thank you also to Daniel Quinlan at the University of Toronto Press for his patience and perseverance.

While this book is an original work, some of the ideas and research appeared in previously published articles, including "*Techne*, Technology and Tragedy," *Techne: Research in Philosophy and Technology* 7, no. 3 (Spring, 2004): 91–112; "Phronesis, Democracy, and Technology," *Canadian Journal of Political Science* 37, no. 4 (2004): 997–1016; "The Politics and Philosophy of the Anti-Science Education," *Techne: Research in Philosophy and Technology* 9, no. 1 (2005): 27–43; "Heidegger's Essentialist Responses to the Challenge of Technology," *Canadian Journal of Political Science* 40, no. 2 (2006): 487–505; and "The Tragic Double Bind of Heidegger's Techne," *Phanex: The Journal of the Existential and Phenomenological Theory and Culture Society* 1, no. 2 (2007): 94–112.

DAVID EDWARD TABACHNICK
North Bay, Ontario
2012

THE GREAT REVERSAL

How We Let Technology Take
Control of the Planet

Introduction
Aristotle's Warning

Imagine two cities. Generations ago, their citizens lived contentedly in the same community until a fundamental disagreement split them apart. The older, unified city was led by a group of legislators who put a priority on family, education, and the common good. They had learned from the wisdom of their ancestors and their own practical experience that these priorities brought with them a happy life. This same city also had a large group of exceptional craftsmen whose task it was to build the schools and monuments the citizens took pride in as well as the roads and docks which allowed for the commerce that helped pay for such civic works. With all of their talent, the craftsmen argued that if less money was spent on public services and more invested in manufactured goods for development and trade, their city could become even more resourceful and prosperous. But the legislators reminded them that the crafts they made helped to support the city and that the city was not there simply to support their crafts. Tired of serving this idea of the good life and fed up with the stubbornness of the legislators, the craftsmen picked up their tools, crossed the river, and started their own city.

For a long time, the two cities on either side of the river thrived in their own ways. The first city kept its modest size and economy, but its outstanding educational system attracted families to come and stay. Its citizens were happy and everyone agreed that it was a good place to live. But they were eventually faced with a major crisis caused by their neighbours across the river. Guided by the vision of those ancient craftsmen, the second city had become a centre of technological innovation, growing very large, with a rapidly expanding economy and high levels of employment. While they did not invest in public services, its citizens had become incredibly wealthy through their skill and industry. To meet the demands of its growing and crowded population, the leaders of the second city used their expertise to reroute the river to support its agricultural and manufacturing sectors, leaving what remained of the water flowing downstream polluted with waste. Unfortunately, this has made life very difficult for the citizens in the first city. Seeing no future in staying, more and more of their young people decided to leave and move to the other side of the river. Their schools half full and suffering from a degraded environment, the first city's legislators now have to decide whether to accept an offer to join the second city that is in the process of overwhelming them. For them, it is more than a choice between two cities, but a choice between two different ways of life.

Of course in the real world we want the best of both cities: tradition, family, education, happiness, and the stability afforded by an agreed standard of the common good as well as innovation, continual growth, and wealth. Yet regrettably, most people today are forced to choose one set of priorities over the other. Unable to escape the overwhelming demands of our modern technological society, it often seems that there is no choice at all. Like the tale of the two imaginary cities, it seems we have experienced a Great Reversal. It is "great" because it is much more than an isolated event experienced in the here and now, but rather is an undercurrent of history running from the origins of Western civilization all the way to the present. It is a "reversal" because we have flipped our priorities, placing the impetuses of technology above our judgments about what makes for a good life. The major consequence of the Great Reversal is the narrowing of human thought and action so that they fit within the confines of our technological society, leaving us unable to think and act upon new ideas that may stand outside of its powerful demands.

We have been aware of a "problem" of technology for a long time. For at least the last two centuries, poets and philosophers have articulated a deepening antipathy to the transformation of both the natural world and the way we live in the wake of the industrial and scientific revolutions, as factories began to crowd our skylines and pollution from smokestacks began to clog our air.[1] But, even before the industrial revolutions or the even earlier scientific revolution there was already available a much older forewarning of the underlying problem of technology. Over two thousand years ago, the ancient Greek philosopher Aristotle realized that human society was determined by two main governing principles, ruling virtues or "directing faculties" through which we could understand the world and our place in it. On the one hand, technical knowledge or *techne* allowed us to build the physical infrastructure of our communities, what the Greeks called the *polis*, and all of the tools and crafts we use in our everyday lives. Through the lens of *techne*, we see the world as something to be worked upon and organized in such a way that it becomes useful to human beings. Trees become lumber, rock becomes stone blocks for building, and animals become food and material for clothes. On the other hand, good judgment or *phronesis* allowed us to pass on and modify the *ethos* or cultural character of the *polis* from generation to generation. Through the lens of *phronesis*, we see the way particular traditions, customs, habits, and laws of a community can be applied to daily decisions, while at the same time we also consider the unique and changing circumstances of current human events. In other words, rather than viewing ethics and politics as products of technical thinking, Aristotle argued that they were instead founded on the always changing judgments of human beings. But, because *phronesis* embraced the diverse and unpredictable practices of human beings, it necessarily lacked the certainty associated with the products of technical knowledge and thus often left the future direction of the *polis* in difficult ambiguity.[2]

From here, Aristotle identified a tension or struggle for supremacy between these two ways of seeing the world. If *techne* were to become the supreme directing faculty that determined the course and character of the *polis*, then life and society could be produced in a predictable and reliable manner in the same way a craftsman produced his crafts. By overcoming the role of chance and the unknown, the technically run *polis* would be safe and secure, eliminating the uncertainty associated with *phronesis*. The problem, Aristotle warned, is that this would also require human beings to be treated as mere material, worked upon and organized so that they too would become predictable, reliable, and useful. It is with this problem in mind that Aristotle decided that *phronesis* rather than *techne* should be the supreme directing faculty of the *polis*. Even though it cannot claim to provide guaranteed results, the *phronesis*-run *polis* would still grant the room that the unique character of human beings needed to flourish. In this city, the citizens would determine the character of its crafts rather than craftsmen determining the character of its citizens.

But now Aristotle's warning has been largely ignored or forgotten. We have chosen to live in the "second city," so to speak. We have accepted the primacy of a technological vision of life and society and have subordinated the role of good judgment. The first and foremost objective of this book is to understand how we arrived at this place, outlining the movement away from Aristotle's warning towards this Great Reversal of the judgmental and technical and why, in its aftermath, we face our present problem of technology. The idea here is to uncover this important but perhaps unfamiliar history beginning with a consideration of Aristotle's articulation of the differences between *phronesis* and *techne* and then highlighting changes in the relationship between these two directing faculties as expressed by later political philosophers. So, while the majority of the following chapters are focused on examining key passages in some of the great works of political philosophy, the larger effort is to highlight the significance of this largely unrecognized history of ideas and its relevance to our contemporary technological dilemma.

Accordingly, the secondary objective of this book is to show how all of this bears out Aristotle's warning that in a technically run *polis* or, put differently, in a technological society human beings become ever more worked upon as mere material and how this kind of treatment may actually bar the practice of *phronesis*. One way to comprehend this is to look at relatively recent breakthroughs in neuroscience and biochemistry that have given us powerful drugs that overcome chronic depression and alleviate anxiety, clearing the way for individuals to live fuller and more complete lives. With increasing skill, we are able to quantify the mind and control its function. As our ability expands, as these drugs become more prevalent in their use and broad in their application, they will no longer be just therapies to treat disease but a means to further control the function of our psychology, enhancing and manipulating our emotional and intellectual states of being. We see a similar possibility in the new field of therapeutic cloning. The discovery of stem cells that can be manipulated to develop into any type of human tissue offers great promise to cure diseases and provide an endless supply of perfectly matched organs for transplant. It is easy to see that, as our ability to quantify the foundational make-up of the human body expands, this therapy will extend beyond medical uses towards a more liberal control of the physical self, augmenting it to make us stronger and faster as well as enhancing our sense organs in any way we desire. In both instances, the original human judgments about what would make us happy and healthy are replaced by an effort to quantify and control what we think and feel. This technologically prescribed ambition may end up subordinating the original non-technological standard derived from our good judgments. Herein lays the disturbing possibility of what can be called a "technological relativism" that leads to a "technological nihilism." If we have no directing faculty outside of technology, then future technological advancement can only be guided by the expansion and refinement of the technical means to quantify and control the world and ourselves directed towards no end in particular, whether happiness or healthiness. In turn, the manipulation of the natural world and human nature may very well transform our minds, our bodies, and the planet into something unrecognizable.

A third objective of the book is to explore the possibility of a "*phronesis* revival" to counter this possible future. The point is that, while we may be on the cusp of a momentous closing off or obstruction of our capacity to judge, there may still also be time to relearn its practice. This "*phronesis* revival" has been promoted in one way or the other by many leading contemporary philosophers and political thinkers. Hans-Georg Gadamer, Hannah Arendt, Jürgen Habermas, and Alasdair MacIntyre, to name a few, all turn to an exploration of ancient Greek philosophy and *phronesis* in particular in an attempt to revive ways of thinking and acting outside of the purely technical.[3] Their work is inspired by a common anxiety that we have become incredibly narrow-minded in the way we live and think – filtering almost all of our thoughts, choices, and decisions through the constricted lens of technology. Because for them the modern world has closed off alternative ways of understanding and living, they have looked to the origins of our civilization – to the philosophers of the ancient world – for guidance and insight, and to revive our flagging sense of community and self. In some part, what is argued in the chapters below attempts to do the same.

As a result, this project might be considered a part of what is called "virtue theory" or "virtue ethics." This is to some degree warranted considering that Aristotle, one of the central figures in the chapters that follow, is a common source and inspiration for many contemporary virtue theorists. MacIntyre's 1981 classic of the field, *After Virtue: A Study in Moral Theory*, for example, is an investigation of the negative consequences of the modern rejection of Aristotelian ethics and a qualified endorsement of their restoration, arguing that "the Aristotelian tradition can be restated in a way that restores intelligibility and rationality to our moral and social attitudes and commitments."[4] Similarly, this book explores whether the revival of the practice of the Aristotelian virtue of *phronesis* can be an effective response to the negative consequences of technology. However, in addition to a more specific focus on the relationship between virtue and technology, what is endeavoured in the subsequent chapters is an attempt to go beyond a history of the development of attitudinal and institutional impediments to the practice of virtue to also include a consideration of the prospect of the impending psychological and physiological obstruction of our capacity to learn and practice virtue. As touched upon above, the increasing effectiveness and popularity of psychopharmaceuticals and the imminent arrival of readily available therapeutic genetic treatments suggests a tipping point where traditional routes to the development of good character will become even more inaccessible and less likely to be retrieved.

This book might also be viewed as participating in a study of the history of political ideas or the history of political philosophy; especially the "historical review" section covering chapters 2 through 5. In general, such studies attempt to highlight prevalent political ideas that link or distinguish historical periods often towards commentary on the composition and character of current politics and society. This approach is well illustrated by Arendt's *The Human Condition*, a wide-ranging inquiry into changing ideas about and articulations of "human activity" from the Athenian *polis* to the politics of the mid-twentieth century. While this present work does not claim a place alongside Arendt's, it does share a similar breadth, spanning the period between ancient Athens to the early twenty-first century, as well as a similar method, in this instance examining the significance of changing ideas about and articulations of the technical and judgmental. And while Arendt, as well as Gadamer and Habermas, have all engaged in serious consideration of the influence of technology on political life within their larger studies, what is attempted in this book might be distinguished as a more particularly political philosophical outline of the origins and evolution of the idea of technology.

In turn, this approach also functions under the notion that prevailing theories of the meaning of the world and the place of human beings in it can have a major effect upon how we actually live. This notion stands in opposition to the environmental determinism that informs Jared Diamond's very popular 1997 book *Guns, Germs and Steel*. Where Diamond argues that civilizations rise and fall almost entirely due to material conditions, such as whether they have a good climate to grow nutritious crops or to domesticate the right kind of animals, the "history of ideas" suggests that intellectual outlook in many ways initiates and directs the course of a society. For example, ideas about human liberty and equality were the inspiration for the American and French Revolutions of the eighteenth century, and it was the later spread of these same ideas that similarly inspired the rise of democratic governments throughout much of the world. So, even though there were no drastic shifts in the climates of or material conditions in these countries, the introduction of the ideas behind democracy nonetheless spurred fundamental change. This is certainly not to say that things like climate or access to nutritious food or some new tool have no role in civilizational change or human development, but instead that they are not in and of themselves the sources of that change or development. The proposition here then is that our *idea* about technology and how it has changed is largely responsible for the creation and growth of our technological society.

Of course, this description also suggests a further alliance with the "philosophy of technology." Prominent philosophers of technology such as Jacques Ellul, Hans Jonas, Langdon Winner, Don Ihde, and Andrew Feenberg engage in specific examinations of the social, cultural, and political origins as well as the philosophical import of technology. A basic disagreement among these thinkers concerns whether technology is an autonomous force or a direct consequence of reigning institutional structures.[5] The first approach is steeped in the idea that technology cannot be controlled or directed by humans because all our institutions and even our ways of thinking are themselves products of technology, leaving us helpless to limit its advance even as it threatens us. The second approach presents "bad" technologies (that cause pollution or ill health, for example) as the outcome of embedded inequalities and injustices within society. Because wealthy and powerful elites control and benefit from the technological infrastructure as it is, there is no impetus to develop "good" technologies. What follows might be considered a third way to think about technology. Rather than viewing it as independent from our influence or simply the tool of the establishment, the argument presented below is that somewhere along the line we began to relinquish the deeply rooted intellectual and practical capacities that allow us to understand and regulate the role of technology in our lives. In turn, we are at present left very diminished in our ability to argue against its advance and feel helpless to prevent its further infiltration into the innermost recesses of our bodies and minds even as it threatens our very humanity. Again, one of the objectives of this book is to explore whether we may still be able to recover and develop these capacities and realize that we still have time to decide how far we want to allow technology to determine our existence and our future.

To sum up then, the main objective of this book is to provide a history of the changing relationship between the judgmental and technical through an analysis of some of the great texts of political philosophy towards understanding how this history is relevant to our present concerns about technology. The second objective is to highlight how our efforts to control the natural world and human nature through the application of technology stem from the decline of judgment and the ascendancy of technical knowledge. And, as just mentioned, the third objective is to explore the possibility of the return of good judgment as a way to limit the role of technology in our lives and understand how our technological society may obstruct or impede this return.

Even though these three objectives are linked by a common or overarching argument, this book can be read in different ways. For readers only interested in problems directly associated with contemporary technology, chapters 1 and 7 are designed as wide-ranging assessments of some of the philosophical, political, and social reasons why we have difficulty in regulating its influence and explore the way in which *phronesis* might be an appropriate response. As first and last chapters, a good part of each is spent either laying out the argument to come or summarizing what has already been presented. Even still, they are relatively self-sufficient and can be read independently of the other chapters. Chapter 6 should be added for those also interested in other responses to the deprivations of technological society, particularly those of Martin Heidegger, as well as the way *phronesis* re-enters the discourse. As mentioned above, the "historical review" section made up of chapters 2 through 5 presents the main "history of ideas" portion of the book, which surveys different articulations of the relationship between the judgmental and technical in the great works of political philosophy. These chapters may be valuable to specialists with an interest in any of these texts and, more obviously, to those looking for what might be called an alternative history of the idea of technology. Importantly, the effort behind these chapters has been to maintain focus on the primary texts and each thinker's discussion of ideas broadly related to good judgment and technical knowledge. While some relevant discussions and interpretations have been provided in notes at the end of each of these chapters, this historical review is not at all an attempt to provide an appraisal of the very extensive scholarly writing available on any one of these political philosophers. Again, the more modest but still, it is hoped, useful attempt is to highlight these thinkers' ideas about the judgmental and technical.

Before moving on, two further notes of caution or clarification are warranted. Outside of its academic roots or relationship to a particular school of thought, this book also provides some observations on the influence of certain technologies. In chapter 1, for example, there is a discussion of reproductive and therapeutic cloning, while in chapter 7 there are brief commentaries on psychopharmaceuticals, steroids, plastic surgery, and gastric bypass surgery. There is no attempt or claim in these sections to offer in-depth assessments of the efficacy or function of these technologies as such. Limited technical details are provided, but only to give a context for how these technologies relate to the larger arguments introduced above. Furthermore, there is also limited analysis of processes and policy decisions on the regulation of particular technologies. So, while the discussion of cloning in chapter 1 looks at some of the controversy surrounding the ban on stem cell research in the United States under the Bush administration, it is not an attempt to offer a comprehensive evaluation or recommendation of specific policies or legislation. Instead, as a work of theory, it is hoped this book will prompt new thinking that may lead to a possible recovery of the practice of some of the virtues upon which our civilization is founded.

1 Finding and Enforcing Limits

As mentioned in the introduction, this chapter is designed as a wide-ranging assessment of some of the philosophical, political, and social reasons why we have difficultly regulating the influence of technology in our lives. What follows then is a broad consideration of our quest to find and enforce limits on technology and why we have such difficulty thinking and acting outside of its terms. It also begins to address the first part of the second objective of the book, showing how in a technological society human beings become ever more worked upon as mere material. To put it more bluntly, the occasion for the writing of this book is the related concern that the increasingly technological character of the world is somehow threatening the future of humanity. After all, technology was supposed to save us from the "nasty, brutish, and short" lives our ancestors lived, but now often seems more like a monster of our own creation that will eventually do us in. Life in the twenty-first century suffers from a collective Frankenstein complex: "I thought that if I could bestow animation upon lifeless matter," Mary Shelley's fictional doctor writes with long overdue regret, "I might in process of time (although I now found it impossible) renew life where death had apparently devoted the body to corruption ... I had desired it with an ardour that far exceeded moderation; but now that I had finished, the beauty of the dream vanished, and breathless horror and disgust filled my heart."[1]

The warning: overcoming the given limits of nature is simply a bad idea because you will get more than you bargained for. But, for reasons just as obvious to us, we are obliged to put tremendous energy, time, and resources into extending life and beating back mortality.

An unwanted death is the most illogical of indignities, the epitome of uselessness. And so we strive ahead with the ultimate conquering of boundaries. Strangely, though, the compelling desire to live a longer, better, more contented life is also the thing that drives forward our de-humanization. The challenge then is to find and enforce limits on new technologies while still satisfying our universal ambition to live well.

It is with this aim in mind that this chapter explores different ways we might meet this challenge. In the process, it also provides a rough sketch or template for the rest of the book. With the exception of the section that immediately follows, the next chapters more or less expand upon the basic divisions of the discussion below. Thus, chapter 2 fo-cuses on ancient Greece, chapter 3 on medieval Christian thought, and chapters 4 and 5 on modern philosophy. Again, these chapters provide a historical review of the changing relationship of "good judgment" and "technical knowledge." Taken together, they show that our pres-ent concerns about technology are not simply a contemporary devel-opment, but are in fact based on a much longer and deeper history. The upshot of all of this is that, where before we had the capacity to judge which technologies would be good to allow into our lives and communities, we have allowed technical knowledge as articulated in today's technologies to overwhelm our capacity to make these kinds of judgments. Again, this change is what is indicated by the phrase "The Great Reversal." Various responses to this difficulty are then ex-plored in chapter 6. For the most part, chapter 6 focuses on Martin Hei-degger's effort to recapture forgotten ways of thinking outside of the now dominant technological mindset and ways of being outside of our all-pervading technological society. Recognizing the impracticality of Heidegger's response, chapter 7 is a call to revive the ancient and per-haps still viable virtue of good judgment as it was articulated originally by the ancient Greek philosopher Aristotle thousands of years ago.

Self-Regulation

Today, we have our own Dr Frankenstein issuing a fresh warning about a new set of technological breakthroughs. But this time, rather than waiting until after the monster springs to life, he is warning us ahead of time. Bill Joy, the former chief scientist and co-founder of Sun Mi-crosystems, wrote a series of articles and gave a number of interviews in an effort to warn the public about the most extraordinary threat we will face from technologies in the twenty-first century. The media blitz began with the provocatively titled cover story in the April 2000 edition of *Wired* magazine, "Why the Future Doesn't Need Us." Joy writes:

> We have yet to come to terms with the fact that the most compelling 21st-century technologies – robotics, genetic engineering, and nanotechnology – pose a different threat than the technologies that have come before. Specifically, robots, engineered organisms, and nanobots share a dangerous amplifying factor: They can self-replicate. A bomb is blown up only once – but one bot can become many, and quickly get out of control.[2]

According to Joy, this coming generation of technologies is far more dangerous than any previous because they are liable to get out of the control of their makers and users. Unlike earlier technologies that are limited by their original design or program, this triumvirate is distinguished by a capability to constantly adapt and grow beyond their intended design. Therefore, he thinks that, even though these technologies may be conceived for peaceful or humanitarian purposes, they present a threat greater than nuclear bombs and other advanced weaponry. Just as Mary Shelley's fictional monster comes back to kill its creator, these technologies may one day destroy humanity.

In his next article, the short but very disturbing "Act Now to Keep New Technologies Out of Destructive Hands," published later that same year, Joy again warns:

> Three new 21st-century technologies – genetic engineering, nanotechnology and robotics (GNR) – are being aggressively pursued by the commercial sector because of their promise to create almost unimaginable wealth. Using them we will be able cure many diseases and extend our lives, eliminate material poverty and grinding physical labor, and heal the Earth. But these new technologies may also pose an even greater danger to humankind than weapons of mass destruction.[3]

He goes onto state that these technologies "pose a large and even mortal danger to our civilization"; that genetic engineering could produce "highly contagious and deadly 'designer pathogens'"; that "out-of-control nanobots could destroy the biosphere"; and that "robotics poses a different threat – the creation of a new life form that may escape our control." The dilemma is clear: these new technologies promise to cure disease but also threaten to generate new, more virulent disease; to heal the earth but also destroy the environment; to eliminate grinding physical labour but also run rampant and out of control. For Joy, any of the benefits these technologies might promise will be far outweighed by their many detriments.

Of course, as the nineteenth century publication of *Frankenstein* can attest, these kinds of warnings are not new. Ever since the first industrial revolution, artistic and intellectual movements have sprung up out of an antipathy to technological and scientific progress. Poets and philosophers reacted against the tremendous social upheaval and environmental destruction they associated with modern society. Yet now we see the "dark satanic mills" of previous centuries slowly giving way to deindustrialized, lower-impact, high-tech economies. Really, instead of destroying us, technology has aided us in myriad ways and, for the most part, these early doomsday scenarios have not played out as predicted. Why should today's concerns about technology be any different?

Still, there are good reasons to take Bill Joy seriously. His warning is not some romantic lament about the loss of a simpler way of life. He is not some handwringing poet or ivory-tower philosopher. As an expert in the development of new technologies, he seems uniquely qualified to warn us about their potential misuse. He has decided that "the only realistic alternative ... is relinquishment: to limit development of the technologies that are too dangerous, by limiting our pursuit of certain kinds of knowledge." He also realizes that this relinquishment, this finding and enforcing of limits on technology, will be no easy task. Because these new technologies offer us so much, it will be extremely hard to walk away from them. In an interview he gave on PBS a few months after the first *Wired* article appeared, Joy tries to explain one way this might be accomplished:

> So we have to somehow prevent everyone from having the ability to create these kinds of massively destructive things in the 21st century, and this is something that scientists and technologists are going to have to take ethical responsibility for. We can't simply release things that are beyond our ability to control, things of such incredible power ... The scientific community has to stand up and recognize that because we're in an information age, we have a different nature of the threat, and we have to take ethical responsibility for the use of the tools, and the information, and the knowledge that we are generating, otherwise we're complicit in their further evil uses.[4]

The exercise of ethical responsibility, self-restraint, and self-regulation by our scientists and technologists could serve as an unassailable limit on the unleashing of these destructive forces. If they decide to keep the genie in the bottle, it will stay there.

Unfortunately, scientists are not in a position to enforce this kind of universal prohibition. The best they can do is warn and advise those who are and can. Take, for example, the role Albert Einstein played in the development of the first atomic bomb. Less than a month before the start of the Second World War, Einstein wrote a letter to President Roosevelt warning him of the dangers of the newly discovered possibility of a "nuclear chain reaction" leading to a Nazi bomb: "This new phenomenon would also lead to the construction of bombs, and it is conceivable – though much less certain – that extremely powerful bombs of a new type may thus be constructed. A single bomb of this type, carried by boat and exploded in a port, might very well destroy the whole port together with some of the surrounding territory."[5]

Because he was in the exceptional position to understand that the Germans were on the cusp of a major breakthrough, Einstein called for "quick action" to secure closer contacts with and provide greater funds to American physicists to help "speed-up" their work on chain reactions in uranium. Despite any ethical reservations about the incredible destructive power of this technology, Einstein, a lifelong pacifist, strongly urged the president to build an atomic bomb and beat the Nazis to the punch. About two months later, Roosevelt wrote back to Einstein with word he had convened a committee to "investigate the possibilities of your suggestion regarding the element of uranium." This committee ultimately gave way to the Manhattan Project and, six years after Einstein had signed his letter to Roosevelt, the dropping of atomic bombs on the Japanese cities of Hiroshima and Nagasaki.

Rather than calling for self-regulation like Joy, Einstein recognized that only a politician would be able to regulate the development of this destructive new technology. From his laboratory, this famous physicist came to see how this kind of research could change the planet and the course of world history. Understandably, without the assistance of people like Einstein, the political leaders of the day could not have possibly foreseen the danger or potential of seemingly innocuous experiments with a little-known element called uranium. The very fact Einstein had to write a letter to Roosevelt to explain the most basic scientific details suggests that the president had no idea of what was to come.

Because science and technology have such a tremendous and growing influence upon contemporary society, we have come to rely on experts like Bill Joy and, in an earlier age, Albert Einstein to help us navigate around unforeseen dangers. We have vaulted our scientists and technologists into a new role of ethical overseers. On the surface, it makes sense that only those who fully understand the complexities of a technology – who know what it can and cannot do – are qualified to make decisions about its larger health and social effects. In turn, legislators and regulators should heed the advice of the technologists and develop and enforce appropriate laws and prohibitions.

Problematically, though, we must then rely on scientists and technologists to make the initial step forward and warn us before the fact. We must ask them to act with ethical responsibility and exercise good judgment when it comes to their own work. Yet, what do they know about such things? While they have expertise in computer science and physics, they do not possess any exceptional knowledge of ethics or experience in making good judgments. As Joy himself points out in the original *Wired* article, scientists may be the last people we should take advice from on such matters. Consider the 1954 testimony of Robert Oppenheimer, the original director of the Manhattan Project: "When you see something that is technically sweet you go ahead and do it and you argue about what to do about it only after you have had your technical success. That is the way it was with the atomic bomb."[6] Obviously, Oppenheimer's disturbing admission about the ethical sensibilities of the Manhattan Project scientists cannot be universally applied to all scientists and technologists. Nevertheless, it is true that ethics is not in and of itself the purpose of science and technology. Many unethical things can still be rightly called scientific and technological. Nazi experiments on concentration-camp prisoners were evil, horrible, and unethical yet still qualify as science. The same point applies to American radiation experiments on military personnel during the Cold War. Oppenheimer himself came to view atomic and nuclear weaponry as unethical but, despite this, the Bomb is still clearly technology. According to Oppenheimer, technical success is the goal of science. Ethics is something else.

The Good and God

But, if not the technologists themselves, who else can find and enforce limits on the development of new destructive technologies? Again, we cannot expect our politicians and legislators by themselves to understand the complexity of the threat. Perhaps in the distant past, before technology became so abundant and complicated, it might have been easier for our political leaders to keep it in check. The ancient Greeks had a profound appreciation for the disruptive influence of technology. While they did not face the unique dangers of genetic engineering, nanotechnology, and robotics, they still understood the need to regulate the role and place of the "technical" or what they called *techne*. Aristotle warned that because technology was itself amoral, it had to be regulated with ethical standards, a strong set of laws, and through wise political decision making based on a concept of a universal and public good. He worried that unregulated technology, free of the guiding influence of these institutions, might be tremendously harmful. Similarly, the Judaeo-Christian tradition presents nature as God's creation, having certain sacred limits that should not be crossed. The Old Testament story of the Tower of Babel can be interpreted as a warning against human arrogance, through technology, attempting to breach the boundary between the heavens and the earth.

Notably, both the ancient Greek and Judaeo-Christian traditions posit supranatural or external standards (i.e., the universal good or God) as guides and limits to human activity. These standards are supposed to help us to make proper choices, create just laws, and live good lives as well as provide clear limits on what we are to build and make. But now we have difficulty accepting these kinds of limitations. For one, because both philosophy and religion claim that these standards have their source in a realm beyond our immediate sensual experience, there is no way for us to rationally verify their precise character or quality.

In fact, it was this same problem that led the earlier modern philosophers to challenge the foundations upon which Western society was built: the classical philosophical traditions of ancient Greece and the medieval Christian traditions of Europe. Because the fixed and external standards both traditions imposed on human behaviour could not be scientifically proven to exist, modern philosophers argued for new, common, and demonstrable standards. They concluded that it was absurd and even dangerous to base ethics and virtue, morality and goodness, law and justice on unproven principles such as some abstract notion of a metaphysical Good or a superstitious belief in the word of God. Thomas Hobbes in his great work of modern political philosophy *Leviathan* (1651) points to the "Invisible Power" of divine right as the source of the widespread civil unrest and revolution sweeping the continent and his native England. For Hobbes, these conflicts stemmed from different religious factions (i.e., Roman Catholics, Anglicans, and Protestants), all claiming authority to govern under God with no way to definitively prove or disprove their right to rule.[7] Instead of the indemonstrable power defended and promoted by religion, he argued that the provable principles of science should be the basis of politics and society. Because the material reality of these principles could be demonstrated through reliable scientific experimentation to anyone with common sense, they could serve as a foundation for universal agreement on the rules and law of society and thus bring peace and security to the world.

But while this new science revealed a corporeal, demonstrable, and common natural order, it was also one devoid of any guiding influence over or set limits on human affairs. Human beings were left rudderless in the swirling sea of the Newtonian universe. God may have set the matter of this universe into motion, but He soon left the scene to allow humans to fend for themselves. According to Hobbes, "There is no such *finis ultimus* (utmost aim) nor *summum bonum* (greatest good) as is spoken of in the books of the old moral philosophers."[8] In other words, there is no external, universal standard by which we can judge the virtue of what we do and what we build. Suddenly, we became profoundly alone with only our internal thoughts and feelings as a guide for such judgments.

This being our lot, the modern philosophers decided that rather than spending time thinking about metaphysics and religion we should turn our attention to human beings. Hobbes's friend Francis Bacon hoped that science would to "some degree subdue and overcome the necessities and miseries of humanity."[9] Likewise, the great French rationalist René Descartes predicted that we were to become the "masters and possessors of nature,"[10] if we only put human thinking in full service to useful applications instead of useless theorizing. Not only was this way of modern thinking to deliver us from poverty, ignorance, inequality, disease, and sickness, it also was to bring an end to the nasty religious wars being fought all over Europe. As the Scottish Enlightenment philosopher David Hume came later to conclude in *A Treatise of Human Nature*, "Reason is, and ought only to be the slave of the passions, and can never pretend to any other office than to serve and obey them."[11] Reason, then, is not for understanding some cosmological good, as the ancient Greeks thought, nor is it for choosing God, as the medieval Christian used to argue. Instead, the human mind is designed to seek out means to satisfy our wants and appetites. Forget about universal love, beauty, wisdom, and truth. The good life must be one of pleasure seeking and pain avoidance or at the very least self-preservation.

Pleasure and Pain

As luck would have it, though, the world seems arranged in such a way as to be opposed and even hostile to the achievement of these goals. Nothing is easy. Nothing is given. We have to work for everything: against the harshness of the elements, the weakness of our own bodies, and the selfishness of our fellow man. As a result, we are compelled to transform the natural world, human nature (our bodies and our minds), and the ways we live with each other to better conform to the internal standards of pleasure and pain, or what Hobbes called our "appetites and aversions." In a sense, this becomes the modern project of technology as manifested in industry, medicine, *as well as* in our ethics and politics.

Today, it seems this project is very near completion. Freed from traditional limitations on human activity, we are able to satisfy our passions restrained only by self-control and the remaining unconquered boundaries of nature. In all of its many forms, contemporary technology is the manifestation of our unchained desires – the set of tools our rational capacity brings into the world to increase our pleasure and decrease our pain. Right now, we are working to build new technologies that overcome the few remaining obstacles to our personal and collective fulfilment, relentlessly transforming the planet and everything on it in the process in the hunt for self-satisfaction.

And so our difficulty in finding and enforcing limits on technology is not due to the new complexity introduced by our particular technologies but instead stems from an earlier dismissal of traditional standards. It was the modern rejection of ancient and medieval limitations that opened the door to the unlimited development of new technology. The problem now of course is that we feel the need for such limits, but find none readily available. If technology is simply the manifestation of our unchained desires, how can we know as individuals and communities what technologies to build, which are good for us and worth pursuing? Just because something gives us pleasure or limits pain, does that also mean it is necessarily good for us? Because the rise of our technological society was precipitated by the dismissal of older standards, we are left with no way to answer these kinds of question and no clear guideposts for the safe introduction of new technologies. In turn, despite our growing reservations about the modern project, we are unable to identify any clear limits for technology beyond the satisfaction of the internal standards of increased pleasure and decreased pain.

The rapid advance of technology freed of traditional limitations has two unsettling and related consequences. The first and more obvious consequence was well illustrated by the unprecedented violence and bloodshed of the twentieth century. As mentioned above, the wide application of science during the previous centuries promised not simply industrial and technological development but also a progressive easing of social ills. Enlightenment thinkers had promised rationally designed political and civil institutions would improve the life of every citizen and bring together the nations of the world to the purpose of everlasting peace. "All these causes of the improvement of the human species, all these means that assure it," the French philosopher Condorcet wrote in 1794, "will by their nature act continuously and acquire a constantly growing momentum."[12] And yet this project also wrought economic depression, famine, genocide, atomic destruction, and world war. Instead of turning our technologies to the sole objective of improving the lives of our fellow man, we developed vast arsenals of advanced weaponry designed to kill them in the most efficient way. Because it does not by itself serve good ends, unleashing technology from its chains also resulted in the satisfaction of the worst of human appetites: for destruction, violence, and power.

The second and less obvious consequence of technology's rapid advance is illustrated by recent discoveries in biotechnology and psychopharmacology mentioned in the opening pages of this book. New cures and therapies for physical and psychological disease have great potential to limit our pain but, at the same time, also push us to redefine the natural make-up of the human body and mind. Rather than simply restoring or repairing, these technologies are also capable of enhancing individuals to be stronger and smarter, possibly giving us more satisfaction and pleasure. While originally conceived to alleviate human suffering, they now introduce a new technologically derived standard by which we evaluate the quality of human health and happiness. As a result, unenhanced individuals may be considered unhealthy and unhappy when compared to those who have received the new treatments. This is problematic because these technologies seem to redefine or recalibrate our passions (i.e., the internal standard of human pleasure and pain). If David Hume was right, this means our rational capacity will be enslaved by these updated passions, which will in turn induce the development of ever-newer technologies that further transform our outward and inner lives.

Hobbes was sure that the common-sense experience of pleasure and pain or the "appetites and aversions" would give individuals an agreeable and inclusive foundation upon which to build a civil society. Now, however, this common foundation is at risk. Our current effort to transform the passions with new technologies is fundamentally altering the composition of the appetites and aversions – the very internal standard upon which we have built our technological civilization. Arguably, if our passions become completely malleable, then we will be left with no common ground. The foundation for our laws, our politics, our culture, and our communities will crumble and fall out from under us into an abyss of relativism.

Again, while we worry about the possibility of such a future, we are also unable to stop it. While we may try to sporadically defend older and outmoded ideas regarding respect for sacred boundaries, ultimately we agree that it is inhumanely cruel to deny the sick and suffering each and every avenue to freedom from pain and disease. To our modern minds, there is little room for fundamental limitations on progress and the forward advance of technology.

Modernity vs. God

When our politicians do attempt to re-inject a supranatural standard into the debate, their arguments must in some way counter the human-centred goals of modernity. A good example of this was on display when scientists and political leaders in the United States first began to seriously weigh the great potential of the new field of stem cell research. If developed, this technology promises to cure previously incurable diseases, provide an endless supply of perfectly matched organs for transplant, and lessen the pain and suffering of millions of Americans. And yet, the federal government baulked at fully supporting the research. Because the harvesting of stem cells required the destruction of a human embryo, the pro-life administration refused to back it. In a 2001 address, then newly elected President George W. Bush explained his position:

> While we must devote enormous energy to conquering disease, it is equally important that we pay attention to the moral concerns raised by the new frontier of human embryo stem cell research. Even the most noble

ends do not justify any means ... My position on these issues is shaped by deeply held beliefs. I'm a strong supporter of science and technology, and believe they have the potential for incredible good – to improve lives, to save life, to conquer disease. Research offers hope that millions of our loved ones may be cured of a disease and rid of their suffering ... And, like all Americans, I have great hope for cures. I also believe human life is a sacred gift from our Creator. I worry about a culture that devalues life, and believe as your President I have an important obligation to foster and encourage respect for life in America and throughout the world.[13]

On the one hand, Bush appreciated the pressing need to alleviate the suffering of those afflicted with diseases such as diabetes, Parkinson's, and Alzheimer's. Stem cells harvested from embryos offer an expedient means to this "noble end." Yet, on the other hand, he also believes in the sacredness of human life, which he argued prohibits the destruction and exploitation of those same embryos. He both accepted the need for stem cell research (i.e., the curing of terrible disease) and derided the consequences of that acceptance (i.e., dehumanization or the devaluing of sacred human life). Problematically, the application of a Christian restriction on technology also requires the rejection of at least part of the modern project to ease human misery. And, because Bush also believes in this project, the government ended up adopting an ambiguous middle-ground policy: actually allowing federal funding of the research to continue but only on existing stem cell lines and offering no restrictions on state-government or private funding.

This angered both opponents and advocates of stem cell research. Opponents argued that it still takes advantage of stem cells that have already been harvested from embryos where, as President Bush says in the same address, "the life and death decision has already been made." It also still permitted the research to go on in private and state-funded labs. From the perspective of these opponents, it is immoral to exploit human embryos for the sake of scientific advancement, regardless of the timeline, previous work, or the source of funds. Of course, researchers and individuals that could benefit from new stem cell–derived therapies were equally upset. They maintain that the United States is being left behind and that there are far too few stem cell lines to engage in effective research. Meanwhile, millions of Americans continue to suffer and die of potentially curable diseases.

Similar conflicts between the goals of modernity and older Christian restrictions are encountered in the debates on birth control, abortion, and euthanasia. Supporters of birth control, for example, view our near-absolute control over fertility and sex as a clear triumph of science over the limitations of our bodies, giving us the ability to better plan careers and families. Pious detractors argue that we should not interfere in our given capacity to procreate no matter the inconvenience it might cause. Likewise, the pro-choice movement argues that women have a fundamental right to decide when and if to have a baby, whereas pro-lifers do not see carrying a pregnancy to term as an option but as an unequivocal moral obligation. Right-to-die advocates contend that the individual should ultimately decide how long to live, whereas opponents understand our time on earth as a proscribed part of a divine plan.

One group says any and all intervention into the human body that gives control over individual existence is allowable. There are no sacred boundaries. In turn, things such as fertility, the cells that make up a fetus, and the body itself are viewed as raw material for manipulation towards the improving, delaying, or ending of life, as the case may be. The other group believes that human life is not infinitely malleable, but bound and determined by God. At least for some Christians, technologies that challenge this notion are evil. While the views on each side are extreme, they have the advantage of clarity. Moderate positions on these issues inevitably get mired in a series of difficult questions about the meaning, start, and end of life. As in the case of stem cell research, the resulting policies and laws that regulate these practices remain frustratingly ambiguous, contradictory, and unsatisfactory: birth control can be used to avoid conception, but not necessarily to end a pregnancy; abortions can be performed up to twenty-four weeks after the start of a pregnancy, but not after; we can end our own lives, but not if we are assisted. These qualifications seem arbitrary. The extremists can override this ambiguity: birth control (whenever, wherever, whatever, or not at all; abortion on demand or none at all (even in the case of rape, genetic disease, or harm to the mother); assisted suicide on demand or no suicide at all, no matter the level of suffering.

Yet, despite any clarity they may exhibit or how progressive or conservative, these solutions are not practical. We can neither accept all new technologies nor prohibit all new technologies. We are forced to choose between them, yet lack the decision-making ability or the good judgment to do so. Our politicians, religious leaders, and scientists seem unable to make a both coherent and humane argument for the limitation of technology.

Embracing Tragedy: Heidegger's Answer

The tension between finding and enforcing limits on technology and the human-centred goals of modernity is also evident in many debates surrounding the introduction of new medical technologies. Are we willing to sacrifice tangible life-saving advances in the name of an abstract notion of dehumanization? Are we willing to let our loved ones die for fear that technology may devalue what it means to be human? For Professor Leon Kass, the first chairman of the President's Council on Bioethics (convened to provide policy advice on the regulation of biotechnology), the answer is yes. Like President Bush, he worries that contemporary technologies have an insidious dehumanizing effect. Furthermore, he sees this process as a symptom of a larger loss of control of the technological project. This was the central point of his testimony in front of the American Bioethics Advisory Commission in 1997 on the matter of human cloning. He warned the commissioners:

> You have been asked to give advice on nothing less than whether human procreation is going to remain human, whether children are going to be made rather than begotten, and whether it is a good thing, humanly speaking, to say yes to the road which leads (at best) to the dehumanized rationality of Brave New World. If I could persuade you of nothing else, it would be this: What we have here is not business as usual, to be fretted about for a while but finally to be given our seal of approval, not least because it appears to be inevitable. Rise to the occasion, address the subject in all its profundity, and advise as if the future of our humanity may hang in the balance.[14]

And Kass continued: "The President has given this Commission a glorious opportunity. In a truly unprecedented way, you can strike a blow for the human control of the technological project, for wisdom, prudence, and human dignity." He made this unequivocal plea to ban cloning not simply because it is dangerous, but because it precipitates the loss of "human control" of technology. According to Kass, the crisis of this loss is not lurking around the corner but is upon us now: we must act immediately towards a radical solution. We must abandon the modern project to "subdue and overcome the necessities and miseries of humanity" and instead accept human mortality and pain. If it were left up to Kass, we would walk away from the promises of stem cell research and cloning as well as steer clear of many well-established life-saving medical innovations. For example, on organ transplants, he writes:

> We have made a start on a road that leads imperceptibly but surely toward a destination that none of us wants to reach ... Yet the first step, overcoming reluctance, was defensible on benevolent and rational grounds: save life using organs no longer useful to their owners and otherwise lost to worms. Now, embarked on the journey, we cannot go back ... There is neither a natural nor a rational place to stop.[15]

He raises similar concerns about other "techniques of prolonging life" such as respirators, cardiac pacemakers, artificial kidneys, and genetic engineering in general.[16] In these warnings, there is less concern about weighing the pain, suffering, and loss of millions against the great need to defend against what he views as an attack on human dignity. He sees that many now common medical techniques are leading to the ascent of a Frankenstein culture where technology gets out of our control, threatening and enslaving us. In turn, even though they may save many lives, these techniques must still be prohibited: we must accept suffering and death as part of being human.

Despite its harshness, this approach seems an effective antithesis to the modern project of alleviating human misery, something Bush's plea for the sacredness of human life was not. Arguably, by sacrificing life-saving and pain-relieving technologies and by accepting misery and embracing tragedy, by expunging our compassion for our fellow human beings and desire to live longer and healthier lives, we might put a brake on the juggernaut of technological advance. And while it seems inhumane and perhaps even irrational, this effort to find and enforce such limitations stems from a desire to defend against dehumanization.

The term "dehumanization" has its contemporary origins in Marxist ideas of manual labourers becoming cogs in the machine of capitalism or industrial society. Kass is not using the term in this sense. He is not saying that the capitalist establishment or industrialists are oppressing and thus dehumanizing a certain class or group of citizens. Instead, he thinks that technology is itself or by itself devaluing the quality of all human life. Inspired by the controversial twentieth-century German philosopher Martin Heidegger, Kass is not primarily concerned about *this or that* technology (e.g., biotechnology or nanotechnology), but accepts a certain truth about *all* technology: it is in its essence a threat to humans. His opposition to the above set of medical technologies is reminiscent of Heidegger's disturbing statement on the character of technology made not long after the Second World War: "Agriculture is now a motorized food industry – in essence the same thing as the manufacture of corpses in the gas chambers and extermination camps, the same thing as the blockading and starvation of nations, the same thing as the manufacture of hydrogen bombs."[17]

For Heidegger, mass agriculture, the gas chambers of the Holocaust, current global politics, and the development of weapons of mass destruction all result from a shared conceptualization of the world and nature as "standing-reserve." That is to say, technology treats all things as stuff to be manipulated. Similarly for Kass, it matters not whether it is organ transplants, prosthetic limbs, or cloning; they all objectify human beings by treating them as malleable matter to be formed in any which way. Heidegger also thought that the same process that manages nature as a resource would come to take up human beings as well. He warns that "man ... comes to the very brink of a precipitous fall; that is, he comes to the point where he himself will have to be taken as standing-reserve."[18] So, just as technology dominates and directs external nature (e.g., trees, animals, the air we breathe), it also infiltrates and manipulates our internal nature: like everything else, human beings become fodder for the spread of technology.

Heidegger explained that this taking up of external and internal nature as standing-reserve is part of what he called the "enframing" character or essence of technology. He did not see this as a recent development, but explains that the original idea of technology was introduced at the very origins of our civilization. He argued that the ancient Greek philosopher Plato's idea of a metaphysical good revealed through human reason put us on a course to transform the world and ourselves to conform to a supranatural standard. From here, we take this idea as a prototype for a perfect world and go about applying it everywhere and to everything, challenging the entirety of existence to bend to its standard. For Heidegger, the Platonic good is not a limitation on technology but its progenitor.

Heidegger and Kass conclude that the only way to find and enforce limits on technology is to reject or get behind this enframing process first articulated in the philosophy of Plato. They ask us to return to a pre-Platonic understanding of technical knowledge or the intellectual virtue of *techne* as a way to reorient or recast our relationship to the things we make. So, rather than simply prohibiting the introduction of new technical products, we must learn to create and relate to our artefacts, all of the things we make, in a profoundly different way: in the same way the ancients produced their crafts. For both men, somewhere along the way we went off the rails, ceding any and all limits on what we make and allowing technology to dominate the planet and humanity. But can we really go back to a time before the rise of the technological idea? Can we really return to a way of life that existed before Western civilization as we know it? Would it really make a difference?

After all, while it seems quite obvious that the meagre crafts of the ancient blacksmith and cobbler are easily distinguished from the computers and genetic engineering of today's technologists, for the most part we still view the difference between these two types of making as *quantitative*. In other words, our present technologies are just more complicated versions of earlier ones. A good example of this quantitative distinction is the claim of anthropologists that such things as the stone chip spears or flints of early humans lay the foundations for later and more complicated technologies. While the amount of technology has increased and become more complicated in our day and age, it is the accumulated result of earlier advances and, as a consequence, future technologies will result from the continuation of that same process. A return to ancient *techne*, then, will make no real difference to our relationship to the things we make.

But Heidegger and Kass argue that there is an essential or *qualitative* difference between *techne* and technology.[19] The big difference is that, where the products of ancient *techne* were temporary impositions of form onto matter, contemporary technology is characterized by an effort at permanent imposition. Adopting Heidegger's language, Kass explains that rather than a "bringing-forth," as is *techne*, technology is "a setting upon, a challenging forth, a demanding made of nature."[20] Where the craftsman "brings forth," works in partnership, or cooperates with the natural characteristics of his materials to construct an artefact, such as a chair or a house, the technologist "challenges" or changes the structure of his materials to make them stronger, more flexible, longer lasting, and so on. For example, a doctor may bring forth the already available health of an individual through a treatment or medicine, whereas cloning, genetic engineering, and organ transplants challenge the natural bounds of the body, creating a wholly new "artefact" that can never return to its previous form, and imposing a permanent new form on the body rather than a temporary one as the doctor did.

It is worth exploring this distinction further. Under normal conditions, because the material of the artefact was still bound by natural characteristics, nature would always "shine through" the imposition of the artist, craftsman, or technician.[21] Another example: a carpenter imposes the form of a chair onto the matter of wood, but once the chair is finished that wood still maintains its natural tendency to rot and decompose in the same way a fallen tree rots and decomposes on the forest floor. In turn, where *techne* still allows a role for nature in its products (including human nature), contemporary technology tries to drive out the movement of nature (e.g., the rotting of wood or the death of the body), doing its best to assure its products will never relapse back into the movement of the natural world.

Because *techne* allows for the maintenance of the natural characteristics of the materials it works upon, Heidegger describes ancient crafts as "scenes of disclosure" that highlight nature. Because the imposition of form is merely temporary in all *techne*, it invites the return of the movement of nature. In this way, the products of *techne* are "scenes of disclosure" in the sense that, through witnessing their destruction (e.g., their rotting away), we come to recognize the temporality of all things, which for Heidegger is the key to living an authentic life in tune with the true finite nature of existence. Presumably, without the building of technical products, this disclosure would not occur. This idea of the "authentic" will be picked up again in chapter 6.

In contrast, contemporary technologies do not cooperate with nature but attempt to replace it. A nuclear engineer can manipulate the structure of natural elements to produce artificial ones. Plutonium is designed to never abide by or return to the characteristics of the uranium from which it was derived. The character of the man-made element plutonium (i.e., its level of radioactivity) is always artificial. Likewise, the genetically altered or "begotten" human is designed to never return to the natural characteristics of the material from which it was derived (e.g., a sick or weak body) and thus is always artificial. In this way, contemporary technological artefacts do not disclose nature but hide it. And, because in a technological society so much of our world is filled with these "undisclosing artefacts," we are cut off from, become unaware of, or forget the essential movement, transience, and limitations of existence and thus live inauthentic lives.

This forgetting also motivates Kass's objections to the above set of life-saving and life-prolonging medical technologies. Because their goal is to do away with pain and delay if not eliminate mortality, he sees them as also eliminating the essential character of humans as limited, finite, and mortal beings. Clearly, by eliminating the limitations imposed by disease, pain, and death we gain more freedom over how we live our lives. But, by the same logic, the limitations imposed by given conceptions of health, pleasure, and happiness must also be eliminated to facilitate that same freedom. Indeed, why should we be bound by human mortality or a particular emotion or state of being? There can be no good or God, no utmost aim or standard whatsoever to guide, direct, or limit our making, including the modern standards of pleasure and pain. At this point, with no standard to guide it, the now postmodern technological project is no longer directed by some antiquated notion of overcoming the necessities and miseries of humanity, but by the singular effort to apply technology to the entire planet and everything on it.

So, while new medical technologies and biotechnologies are merely a specific articulation of the much larger development of technology, they do represent an important turning point. Because they challenge the internal standards of pleasure and pain that served as the foundation of the modern project, these technologies remove the single remaining guidepost by which we judge the safe introduction of new technologies into our communities. Our growing inability to find and enforce limits on technology is a reflection of this technology-driven nihilism.

For Heidegger and Kass, a limit on technology can be found in a return to the tragic. Yet, while embracing tragedy and returning to a more primal relationship with our artefacts may stem the onset of technological nihilism, it seems more than unlikely that most of us would voluntarily select this option. Faced with a choice between saving ourselves or our loved ones from a painful death and stemming the tide of dehumanization, even the most ardent critics of technology would probably choose the former. Yes, there are certain religious groups that place various restrictions on the use of specific technologies and medical treatments: for instance, in refusing to allow their children life-saving blood transfusions. Breaching the sacred boundaries between life and death is a greater evil to them than allowing their sons and daughters to suffer and die. However, as the modern philosophers long ago realized, there is little hope that we can reach an agreement on where these sacred boundaries lie, and thus it remains impractical (in a free society) and cruel to use them as guides for the prohibition of new technologies.

Politics: Aristotle's Answer

For Kass, because individuals will not themselves voluntarily reject these "challenging" medical technologies, the only way to find and *enforce* limits is through a politically derived prohibition. This is why he participates in the political process and argues that there is still the possibility to assert human control over technology through the legislative process. Clearly, he does not trust the scientist and technologist to self-regulate nor is he confident that individual religious conviction is enough to keep these technologies in check. Rather than deferring to the experts or invoking a supranatural or religious standard to justify such prohibitions, our political leaders must do a much better job explaining how and why these new technologies represent such a tremendous danger that we cannot allow them into our communities.

This is Francis Fukuyama's central point in his 2002 book *Our Posthuman Future: Consequences of the Biotechnology Revolution*. Fukuyama, also a member of the President's Council on Bioethics, argues that politicians and legislators must pay more attention to new developments in technology and, more specifically, biotechnology. He writes that "countries must regulate the development and use of technology politically, setting up institutions that will discriminate between those technological advances that promote human flourishing, and those that pose a threat to human dignity and well being."[22] As he notes, it is wrongheaded to think that we are simply helpless to stop the advance of technology. We already accept a role for politics in its regulation and have wisely and successfully kept certain technologies under strict control. Countries do not allow the free development and use of nuclear weaponry, for example. This being the case, we can also regulate new biotechnologies.

He also introduces, similarly to Kass, a set of standards by which we can judge the virtues of technology: if the technology promotes human flourishing, human dignity, and well-being. But, it can immediately be asked, who gets to decide what promotes and what threatens these things? Fukuyama explains:

> The case that I will lay out here might be called Aristotelian, not because I am appealing to Aristotle's authority as a philosopher, but because I take his mode of rational philosophical argument about politics and nature as a model for what I hope to accomplish ... Aristotle argued, in effect, that human notions of right and wrong – what we today call human rights – were ultimately based on human nature. That is, without understanding how natural desires, purposes, traits, and behaviors fit together into a human whole, we cannot understand human ends or make judgments about right and wrong, good and bad, just and unjust.[23]

So, rather than returning to a time before Plato, he sees an answer in the period right after, in the work of Plato's greatest student, Aristotle. His "case" is Aristotelian because the ancient Greek philosopher Aristotle argued that the politician or statesman best exemplifies an ethical understanding of human ends, between right and wrong, good and bad, just and unjust, and so on. What is more, unlike the technologist, a political leader also has the capacity to enact and enforce these decisions through the legitimate coercive power of the state.

Rather than a return to ancient *techne*, this is an entreaty to a different ancient intellectual virtue: *phronesis* or good political judgment. While today we might argue that the politician is the last person we should look to for a restoration of ethical responsibility, in the Aristotelian model politics necessarily implies ethics, and the statesman necessarily implies a person of good and ethical judgment. For Aristotle, to be "ethical" was more than simply knowing right from wrong, but also meant the capacity to act upon that knowledge. In both his *Politics* and *Ethics*, Aristotle argued that the statesman, the person with *phronesis* or the *phronimos* knows "what is the end or aim to which a good life is directed"[24] and was the most able to articulate that knowledge into good laws and policies, the practice of politics being the "supreme directing faculty" that guided all other good action or good decisions in the city.[25] The political was the lens through which the *ethos* or character of a community passed, not only to the present citizenry of a community but also to the next generation of citizens. Tellingly, the word *ethos*, from which we get the word ethics, was originally used to describe the shared ancestral den or burrow of animals. In this sense, ethics is as much about upholding the long-established conventions of our forbearers as it is about making the right decision in the here and now. The role of the *phronimos* was more than simply securing the safety and security of the citizenry, but also required the upholding of time-tested traditions, customs, and habits proven to bring with them a good life.

So, Aristotle's statesman-*phronimos* is not a cobbler, a blacksmith, or a house builder, but still understands the highest end to which all these technical crafts aspire. Likewise, Fukuyama's political leader may not be a geneticist, biochemist, or roboticist, but can still judge whether these technologies are directed towards good ends. Because the statesman or *phronimos* knows what are the good ends of their political community, they are in the best position to judge what crafts or technologies belong within its walls so to speak, not the craftsman, technologist, or religious leader. For Fukuyama, our technological age demands our political leaders be recast in the role of the Aristotelian *phronimos*, attempting to find a middle way through contentious debates on technologies and also exercising the power of a statesman to enforce their decisions. Therefore, a great onus is placed upon the judgment of our politicians to find the right balance between human flourishing and potential affronts to human dignity.

Importantly, Aristotle's *phronimos* does not hold a fundamental suspicion against technology. The point for Aristotle is that technology is good only when subordinated by higher virtues such as those associated with ethics and politics. He is clear that we need the products of *techne* in order to live good and full lives,[26] but also writes that "it is for the sake of the soul that these other things [external goods] are desirable, and should accordingly be desired by every man of good sense – not the soul for the sake of them."[27] This is a warning that external goods, the products of technology, should be used in the *service* of being a good person and living a good life. In other words, there is a hierarchy of goods or virtues that makes the higher end of politics the guiding principle of the lower ends of technical knowledge. For Aristotle, because politics has a higher end, it determines the lower ends of technology rather than the other way around. In turn, *phronesis* represents another alternative to the modern project. It suggests that by focusing solely on satisfying pleasures and avoiding pain (by basing modern society on this universal standard), we have amplified one aspect of humanity into the one and only characteristic of being human. But, because *phronesis* subordinates the individual appetites and aversions, or pleasures and pains, to a higher community-wide goal, it becomes possible to prohibit new technologies even though they may alleviate suffering and save lives, giving us bodily satisfaction.

So, although stem cell research may derive therapies with benefits to the individual, and while it may have some virtue or some good, those benefits must be considered in the context of the greater goods of the political community. In fact, the ethical and legislative debates surrounding stem cell research suggest that it is an ideal "case study" of how we might apply *phronesis* to our contemporary concerns about specific technologies. The idea that the common good must help inform the realization of individual goods inspired the President's Council on Bioethics' first major report in July 2002, *Human Cloning and Human Dignity: An Ethical Inquiry*. In this document, the council (including members Kass and Fukuyama) presented recommendations for two types of cloning: reproductive cloning and therapeutic cloning. The council expressed unanimous opposition to reproductive cloning or, what they called, "cloning-to-produce-children." While they recognized some of the potential merits of the technology,[28] they decided that, when considered within the larger context of society, any potentially good outcomes would be far outweighed by wider negative impacts.[29]

Overall, the council's opposition to reproductive cloning did not focus so much on the act of cloning itself as on the problems that may arise post-cloning. Questions of long-term health, freedom, family, identity, and society were at the forefront. The council recognized that the cloning debate cannot remain solely focused on technical and safety issues, but must also consider the larger societal effects of the technology. Robert Wachbroit, a research scholar at the Institute for Philosophy and Public Policy in Maryland, sums this idea up succinctly when he writes: "The ethical issues of greatest importance in the cloning debate … do not involve possible failures of cloning technology, but rather the consequences of its success."[30] According to this thinking, the real problems of cloning are philosophical, ethical, political, and social and cannot be addressed on only scientific grounds. The implication is that in order to understand fully the impact of technologies like cloning we must go beyond the expertise of scientists and technologists and include ethicists, philosophers, sociologists, and others to advise on larger issues and dilemmas. Following this logic, membership on President's Council was divided among a bioethicist, a political philosopher, a professor of Christian ethics, a neoconservative columnist, a professor of metaphysics, and other distinguished philosophers, law professors, *as well as* medical doctors, biochemists, and neuroscientists. And, despite this diversity, they all agreed that reproductive cloning should be banned.

However, the council was split on whether to allow therapeutic cloning or therapies derived from what President Bush called human embryo stem cell research. In a 10 to 7 decision, they recommended that the original moratorium on federal funding be extended for four more years.[31] In contrast to the thinking that went into their reproductive cloning decision, the council now focused on the harvesting of stem cells rather than the post-cloning social and political implications. The main issue for many council members was that the cloned embryos must be destroyed in order to harvest stem cells. In other words, the council split on the contentious subject of the rights of the unborn. Therefore, where they thought through the long-term, community-wide, post-cloning implications of reproductive cloning, there was less consideration of the societal effects of genetic treatments, therapies, and enhancements derived from therapeutic cloning and stem cell research. The debate remained centred on the status of the human embryo.[32] The first decision used the Aristotelian political model, whereas the second decision employed the "sacred boundaries" argument. Significantly, even though reproductive cloning might be considered to have even more potential medical benefits than therapeutic cloning, its prohibition has received almost universal domestic and international acceptance. Yet, because the therapeutic cloning ban is based on a contentious religious division, it remains the subject of widespread controversy and indecision.

This is just one case or example of the way in which *phronesis* might help us find and enforce limits on technology. As will be illustrated next in chapter 2 and discussed more later on, *phronesis* is also a rebuke of Heidegger's enframing argument, which identifies "the good" as the progenitor of technology. Because this virtue is based on the particular experience of living in a community, it does not treat "the good" as a supranatural standard divorced from existent realities. Rather than providing a monolithic prototype for the transformation of the world, *phronesis* balances larger considerations of what is good with the particular character of individuals and communities. The *phronimos* admits that what works in one instance might not work in another, refusing to adopt a one-size-fits-all approach to decision making. Of course, it is this very "case-by-case," "off-the-cuff" quality of *phronesis* that also makes it the subject of criticism. Indeed, even in Aristotle's day, it was argued that politics must be more reliable and predictable, treated as a product of technical knowledge rather than the practice of good political judgment.

A Brief Note on the Historical Review

The next four chapters are devoted to a historical review of the judgmental and the technical as they are articulated in some of the "great books" of political philosophy, in good part fulfilling the main objective of the book to outline "the movement away from Aristotle's warning towards this Great Reversal of the judgmental and technical," as stated in the introduction. The plan for these chapters is to pick out relevant passages of representative or significant texts and reassemble them in an outline that catalogues different expressions of these concepts as well as the changing relationship between them. So, while the Greek word *phronesis* does not mean exactly the same thing as the Latin *prudentia*, the Italian *prudenza*, or the English word prudence, it is possible to see that they all describe the capacity for judgment. Likewise, while Plato's "kingly *techne*," Hobbes's "political science," and Condorcet's "social mathematics" are different ideas, they all describe the technical control of human beings. In the broad scope of things, this outline will not only demonstrate changing ideas about the judgmental and technical but will also highlight the interrelationship between an overall decline in confidence in our capacity to judge and our continually increasing dependence on technical control.[1]

Suffice it to say, just because this outline highlights the merits of Aristotle's discussion of *phronesis*, it should not be taken as an atavistic or sweeping endorsement of ancient Greek society, either as Aristotle described it or how it was actually lived. Rather than a romantic celebration of the past or a call for the resurrection of the Athenian *polis*, the aim of these chapters is to show how our concepts of judgment and technology have reversed and how this reversal has contributed to our present problem of technology.

2   *Phronesis* vs. *Techne*

This chapter is a discussion of the division of *phronesis* and *techne* as well as a comparison of Aristotle's call for "*phronetic* rule" with Plato's alternative call for "kingly *techne*" as presented in his dialogue *The Statesman*. A key objective of this chapter is to show the way *phronesis* serves as a bridge between our everyday experience of the good life and our intellectual understanding of what is good and right. Critically, the practice of this ancient virtue is rewarded with the experience of happiness as manifested in physical pleasure, material wealth as well as the respect of our peers. As will be discussed in chapter 3, the breaking of this connection between our everyday experiences of happiness and our ethical understanding of what is good and right, lends itself to a new effort to impose an otherworldly order onto the natural world and human nature.

In his *Ethics*, Aristotle described *phronesis* as "a reasoned and true state of capacity to act with regard to human goods."[1] Like *techne*, *phronesis* is an "intellectual virtue" or a personal quality that helps guide individuals to correct thinking. These virtues, which also include *episteme* (empirical knowledge), *sophia* (philosophical wisdom), and *nous* (comprehension of the first principles), are developed and honed through a good education and individual contemplation. The other set of virtues identified by Aristotle, the "ethical virtues" such as courage, moderation, and generosity, while still connected to the intellectual virtues, have less to do with thinking and relate more to controlling the emotions, passions, and appetites. He described them as good habits that are (hopefully) learned at a young age in the home. Critically, the learning and performance of the ethical virtues is guided by and dependent upon the practice of the intellectual virtue of *phronesis*.

Aristotle explained that all these virtues fall within a larger hierarchy, with the lower virtues guided by the influence of the higher virtues. For example, in a city at war, the bravery of a soldier defending the community from attack is superior to the skill of a cook able to prepare good food. While both citizens could be considered virtuous, the needs of the soldier to eat healthy rations so he is ready and able for battle should determine what kind of food the cook ends up making. Here, the virtue of the soldier should guide the cook and the skill of the cook should serve the soldier.

Aristotle took this same idea and went on to advise that in general the most intellectually and ethically virtuous citizens should rise to positions of power and prestige, the less virtuous serving them, and that the least virtuous or most vicious should be shunned or imprisoned. He also warned that if this hierarchy is upset or disrupted, then the city will be dysfunctional and unable to provide essential goods for its citizens. And while the ancient world has numerous examples of city-states and political leaders not heeding this warning or following this hierarchy, it is still fair to say that the practice of virtue was the foundation of ancient Greek politics, law, ethics, and social norms.

Notably, where *techne* and *phronesis* are both associated with the constantly changing and practical world of human affairs, the other three intellectual virtues identified by Aristotle are instead purely theoretical or connected to things that do not change. *Episteme*, for example, relates to the study of the natural world. Employing this virtue, a theoretical scientist might study astronomy or the moon's orbit. Obviously, this study is not going to change the position of the stars or the way the moon moves around the earth. The effort is not to change but merely to understand. So, while this kind of empirical knowledge of the heavens can be demonstrated or proved through a scientific account, it does not have any immediate practical or useful application. We might call "pure sciences" such as contemporary astrophysics or the study of quantum mechanics good examples of *episteme*. Likewise, employing *sophia*, a philosopher may seek out the truths of the cosmos. However, if he were to grasp them, he has no expectation of changing those truths or interest in using that knowledge towards any useful end. Socrates, the model philosopher of the ancient world, was not exactly the most productive or practical fellow, spending his days wandering the streets of Athens looking to start an argument while paying no attention to his job or his family. *Nous* is an even more abstract, impractical virtue that involves something like an intellectual, almost spiritual grasp of what Aristotle called the "un-moved mover" or the very origin of existence.

All told, because these intellectual virtues do not really change the world, people such as theoretical scientists and philosophers have only an indirect interest in the goings on, problems, and concerns of their fellow citizens. Indeed, these thinkers might be said to have their heads in the clouds, craned upward to the sky (or downward to their microscopes) so that they can better contemplate the outer and inner workings and meaning of the unchanging world. The kind of thinking associated with these three theoretical virtues might be best described as "an end in itself" – personally fulfilling, but not expected to provide anything of use. These virtues should instead be judged on their own merits, not whether they produce some sort of new gadget, timesaving device, or moneymaking scheme.

Of course, astronomers, astrophysicists, and even philosophers may object to being labelled as "useless." Funnily enough, Aristotle coun-tered this very concern with a story about the philosopher Thales. As the story goes, Thales was reproached by his fellow citizens for his lack of wealth, told that he spent far too much time studying the movement of the heavens and not enough time making money. To counter his critics, he used his observations of the stars to predict the coming of a good growing season and a bumper crop of olives. Keeping the details of his forecast to himself, he bought up all of the olive presses at a low price. When the large harvest of olives came in, he then rented the now highly valued presses out at a large profit. This proves, Aristotle ex-plained, "that it is easy for philosophers to become rich if they so desire, though it is not the business which they are really about."[2] The point of the story is that, because the practice of philosophy is hierarchically su-perior to the lower virtue of moneymaking, the philosopher is actually able to understand economics and, in turn, influence its practice. Of course, the real business of philosophy is instead to help us understand and explain fundamental truths and goods, as hard as this may be.

Those with the virtues of *techne* and *phronesis*, by contrast, change the world; they directly participate in and improve the lives of their neighbours and fellow citizens, solving problems and even making money. As mentioned in chapter 1, *techne* "brings into being" the "ex-ternal goods" everyone requires to live a comfortable life, while *phro-nesis* guides the development of good ethics and laws, including the regulation of the amount and kind of external goods or technical prod-ucts allowed in their political community or the *polis*. *Phronesis* is not opposed to or in conflict with *techne*, but limits and guides its role. So, rather than allowing the craftsmen themselves to develop appropriate laws and prohibitions on their various arts, a good politician informed by *phronesis* must have a capacity to decide which technical products will serve the city and which will harm it.

This relationship between *techne* and *phronesis* is part of the hierarchy of virtues that leads the *polis* as a whole towards the final goal of the "good life." Aristotle named this goal *eudaimonia*, which can be translated as felicity, happiness, or, more generally, living well. For Aristotle, living well meant that a person is fully and completely satisfied in all aspects of his life – having personal health, a sense of public purpose, as well as intellectual fulfilment. He also recognized that different people may require different kinds and ratios of satisfaction; some focusing more on health, others on wealth, and yet others on intellectual pursuits such as philosophy. Likewise, different cities required different composition of citizens, some having a greater need for artisans, others soldiers, and yet others politicians.

As Aristotle concluded, persons travelling down their proper path to the good life, fulfilling their purpose – their *telos* – will experience great pleasure and happiness. It follows that one can know they are travelling down the wrong path if they are in pain (physically, emotionally, or intellectually) and miserable. How one *polis*, one set of ethics and laws, can facilitate and provide for as well as educate and direct the multiplicity of paths to *eudaimonia*, allowing for all citizens to reach their final goal, is the critical and difficult task given to community and political leaders. If these leaders are successful, not only will each citizen be happy but also the *polis* as a whole will be considered a pleasant and good place to live.

Notably, Aristotle did not think there can be *eudaimonia* for people living outside of the *polis* because "man is by nature an animal intended to live in a *polis*."[3] For him, we are political animals (*politikon zoion*) because our *telos* can only be attained within the context of the comforts, protection, and education found in a well-run political community. Obviously, a community racked by inferior leadership, political corruption and ineptitude, poverty and violence will be unable to achieve this most important goal.

Techne and the *Technites*

So, according to Aristotle, *techne* and *phronesis* are key virtues for the proper running of the *polis*. While the moon will continue to orbit the earth and the eternal truths of the cosmos will continue to be true with or without scientists and philosophers to study and think about them, the amount and kind of external goods necessary for living well will simply not exist without the technician or *technites*. They have a precise know-how to make things that people need: the cobbler has a *techne* of shoemaking; the house-builder has a *techne* of house building, and so on. The distinction of *techne* is that it produces useful things in a predictable and consistent manner. Rather than relying on things produced by nature (*physis*) or waiting for things to come by chance (*tuche*), as is the fate of wild animals, *techne* gives human beings the tremendous advantage of making what we need when we need it. The craftsman's art satisfies our material needs and frees up time for more lofty concerns and practices such as politics and philosophy. Clearly, without citizens with this sort of expertise, these valuable products would simply not appear in the *polis*, leaving citizens without shelter, shoes, and agriculture as well as all the other technical products we associate with living well.

According to Aristotle, *techne* is "a state of capacity to make, involving a true course of reasoning."[4] Significantly, because it involves "a true course of reasoning," the *technites* is able to clearly articulate and teach that expertise to students and apprentices. In turn, not only is the virtue of technical knowledge easily passed down from one generation of technicians to the next or transferred from one civilization to the next, it can also accumulate and incorporate new skills and expertise as they are discovered and refined. The wonder of *techne* is that it can be amassed, written down in an instruction manual, appended and updated, growing larger and more complicated with every passing day.

Phronesis and the *Phronimos*

Like *techne*, the intellectual virtue of *phronesis* also relates to the mutable world of human beings. While the term is found throughout many ancient Greek texts, it is hard to define because it does not have an obvious contemporary counterpart, as is the case with *techne*. Sometimes it is called prudence. Other times it is practical wisdom, practical intelligence, or practical deliberation. A better and simple translation is "good judgment" because it suggests that *phronesis* requires both intelligence *and* experience. Again, Aristotle explains that *phronesis* is "a reasoned and true state of capacity to act with regard to human goods."[5] In other words, simply having knowledge about what the right thing to do is is not enough. Ultimately, *phronesis* is characterized by acting on that knowledge. A leading scholar on the subject explains: "If I *see* what the situation requires, but am unable to bring myself to act in a manner befitting my understanding, I possess judgment but not *phronesis*."[6] Merely knowing what the right thing to do is not *phronesis* unless it is swiftly and intuitively followed by right action.

What is important here is that *phronesis* is not merely "knowledge," as is true with *techne*, but describes the good action of a particular kind of person: the *phronimos*. It is "good" action because it is informed by a "regard to human goods." Regrettably, there is no simple or straightforward way to know what a human good is. We cannot turn to an instruction manual of human goods or take a course on how to act with regard to human goods. Unlike a *technites* teaching an apprentice, the *phronimos* cannot successfully instruct somehow on how to make good decisions. Because no two decisions can be based on the exact same circumstances, the criteria of a choice to do the right thing are constantly changing. Despite this difficulty, a *phronimos* does not freeze up when a bold move is required; *phronimoi* are not paralysed by indecision, doubt, or fear but are quick on their feet, confidently rising to any occasion and acting towards a good result.

Aristotle, for one, pointed to Pericles, the cool-headed and brilliant saviour of Athens during the Peloponnesian War, as a good example of a *phronimos*. However, it is Thucydides, the great historian of that same war, who provided perhaps the best description of a *phronimos* in this passage on the unrivalled military and political leadership of Themistocles:

> Indeed, Themistocles was a man who showed an unmistakable natural genius; in this respect he was quite exceptional, and beyond all others deserves our admiration. Without studying a subject in advance or deliberating over it later, but using simply the intelligence that was his by nature, he had the power to reach the right conclusion in matters that have to be settled on the spur of the moment and do not admit of long discussions, and in estimating what was likely to happen, his forecasts of the future were always more reliable than those of others ... To sum him up in a few words, it may be said that through force of genius and by rapidity of action this man was supreme at doing precisely the right thing at precisely the right moment.[7]

The word Thucydides used to describe Themistocles's natural genius is *synesis* – often translated as "practical intelligence" and also a common way to translate *phronesis*.[8] Suffice it to say, while General Themistocles may have been born with a natural propensity for quick thinking, it was his many experiences on and off the battlefield that gave him the competence to make good decisions and allowed him to become such an admired figure. According to Aristotle, one can only become a *phronimos* through a process of trial and error, because good judgment is not innate but a good habit (*hexis*) acquired through practice.[9] This is why Aristotle thinks young men may be clever or smart but not wise in practical matters[10] – they simply lack the required experience.[11]

By acting towards and achieving good things repeatedly (and avoiding bad things), a person of good judgment is in time able to rationally understand or perceive the common character or quality of human goods and act accordingly in an ethical manner. Eventually, the *phronimos* can "deliberate well about what is good and expedient for himself, not in some particular respect, e.g. about what sorts of thing conduce to health or to strength, but about what sorts of thing conduce to the good life in general."[12]

The most critical step to acquiring the habit of *phronesis* is following the example of a pre-existing *phronimos*.[13] By way of imitation of these role models, a child or a student will get into the habit of acting in a good and ethical manner. In turn, *phronesis* can only be developed through a life lived in a city filled with good parents, friends, educators, and political leaders – the *phronimos* is a reflection of, comes from or out of, the community. For example, while a young child might at first have to be made to eat healthy food (or "habituated" as Aristotle would put it), as children get older they will come to accept and understand the "good" or the virtue of healthy eating. By imitating the temperate acts of his parents, a child will eventually develop a mastery over his passions, appetites, and desires.

At first, there may be considerable tension between this virtuous action and, let us say, the child's excessive appetite. But, with enough practice, this tension will dissipate, the child's soul will be ordered (so to speak), and the ethical virtue of moderation will help him make good decisions about eating. Importantly for Aristotle, we should not repress natural inclinations but develop and guide them to good ends. So, a child's appetite for food or, when he gets older, passion for sex are the initial spurs for later self-understanding and development. This child will work their way through the lower goods associated with health and the body towards higher goods guided by that which is pleasant and the example of a *phronimos* or role model. Through proper choices, the ethical weakness, immaturity, or intemperance that might instead have led them to gluttony or lust later in life are pushed aside. Again, for Aristotle, the experience of satisfying the lower appetites, seeking that which is pleasurable and avoiding that which is painful, is a critical first step towards being able to make good judgments about higher and more important things.[14] What might begin as individual decisions about health will eventually lead to good decisions about "life in general" that might later be applied to more complicated matters or higher goods such as politics and law. In other words, at a certain point a child will be able to do more than simply mimic the behaviour of the *phronimos*, but will apply the good habit of *phronesis* to the unique circumstances she faces in her daily life, demonstrating her own good judgment and ethical virtue and thus becoming a responsible citizen in her community.

Needless to say, a child who has gluttonous parents and never learns how to take care of his or her own health, family, or household is an unlikely candidate to be a good citizen, political leader, legislator, or ethical exemplar. The problem, as Aristotle saw it, was that many societies had lost this critical connection to the "human goods." He pointed to the "vulgar decline" of statesmen who are concerned only with the "useful" and "profitable" as well as empire building.[15] Critically, if the *polis* fails to pass down the bases of its *ethos* of the good life to the next generation of citizens, then the constitution of the whole city will become deviant. As a result, the next generation of citizens brought up by that city's elders, parents, and legislators will be unhappy.[16] Arguably, once the link between generations is broken, recovery of "human goods" and "the good life" is difficult if not impossible. Indeed, this is what contemporary "*phronesis* revivalists" like Gadamer, Arendt, and MacIntyre think has happened to today's technological society and why they seek to revive the ancient practice of *phronesis* .

In a sense, the *phronimos* serves as the bridge or link between generations. Because he accumulates an understanding of "human goods" in general, he is also able to ensure that the fundamental tenets of the good life are always present, not overturned in the clamour of change. He knows "what is the end or aim to which a good life is directed"[17] and "must labour to ensure that his citizens become good men."[18] He possesses an understanding of the *polis* beyond the mere conventions of city life, understanding how to live a good life in general. *Phronetic* leadership requires both a political education (what Aristotle called "political science") on how politics works in all communities and a practical knowledge of one's specific community (e.g., its terrain, neighbouring communities, the size and make-up of its population, and its legislation). Through a combination of larger considerations of politics and law, political education, and political experience, the *phronimos* is able to strike a balance between the particular needs of his community and what is good for all communities.[19] The *phronimos* realizes that the particulars of life in the city are always changing or growing and yet the city's overall tenor remains the same.

With this explanation in hand, we can begin to understand how the contemporary revival of the ancient virtue of *phronesis* might help us find and enforce limits on technology. Because it includes a consideration of overarching human goods beyond mere bodily satisfaction, *phronesis* gives us access to higher virtues by which we might judge what technologies we should allow into our communities and those we should prohibit. Put differently, just because we can produce something that satisfies some need or desire does not mean that it should be produced. Arguably, with a *phronimos* or a person with good judgment at the helm, we will no longer be entranced by the narrow promise of technology to alleviate our pain or enhance our pleasure, but will instead regulate the satisfaction of these lower goods in subordination to the higher virtues of the good life in general.

As it was put forward in the last chapter, the difficulty with this way of finding and enforcing limits on technology is that individuals will not, by themselves, voluntarily reject life-saving, pain-reducing, or pleasure-enhancing technologies. In turn, it may be that the only way to find and enforce limits on technology is through a politically derived prohibition. Needless to say, this may seem at first glance profoundly undemocratic. With this objection kept firmly in mind, an alternative to the *phronimos* might now be explored.

Kingly *Techne* and the Craftsman-King

Not everyone in the ancient world agreed with Aristotle's idea that the *phronimos* was the best political leader or that he would be the most able to deliver the good life. Aristotle's teacher Plato seemed to recommend that the *technites* would be a far better candidate. Plato appeared to argue that, if a technician were in charge, he would be able transform the city and its citizens to be precise and predictable in the same way a craftsman makes his crafts. Rather than simply accepting that the thoughts and actions of human beings are unpredictable and impossible to perfectly anticipate, Plato thought that, under the controlled conditions of a technically designed and run *polis*, we could have a guarantee of prosperity and happiness in the same way a craftsman could guarantee the qualities of his crafts. This led Plato to explore the possibilities of what he called "kingly *techne*." A techno-political rule, he theorized, might allow for the control, stability, and dependability that would give a political leader the unwavering ability to direct and mould a city and its citizens in the same way a blacksmith forges horseshoes and swords. For Plato, rather than the practical and spontaneous decisions of the *phronimos*, the *polis* would be better served by the productive and predictable expertise of the technician.

While this is a theme in many of the Platonic texts, it is on clearest display in *The Statesman*. In this strange and striking dialogue, Plato blurred the roles of the craftsman and the politician, writing not about the infamous philosopher-king of *The Republic* but instead the possibilities of a craftsman-king practising a kind of politics comparable to the crafts of carpentry, shepherding, and weaving – a "science of government" that treats citizens like material no different than wood, sheep, or wool. Just as craftsmen transform these basic materials into their crafts, the craftsman-king can transform people and territory into a good and healthy city. Here, *techne* is no longer limited to producing basic external goods, but is applied directly to the high art of politics.

Of course, Plato admitted that the job of the craftsman-king is quite a bit more challenging and complicated than that of the ordinary technician. He explained, for instance, that where a carpenter builds with inanimate materials such as wood, a statesman "has a nobler function, which is the management and control of living beings."[20] And, because "living beings" are a far more complex material than wood, the statesman might better be compared to a shepherd who, instead of herding sheep, is a "herdsman of humanity." But Plato also conceded that the kingly-technician faces yet a further complication. Where sheep do not generally give advice to the shepherd on how or where they should be herded, human beings have many opinions on how their city should be run and under what rules and laws they wish to be governed. Unfortunately, the citizen-sheep are often too obstinate, under the flawed impression that they know how to run the city better than the shepherd. For Plato, this bedevilling and all-too-common lack of diffidence to authority was something the shepherd should and could breed out of his flock. Like a weaver of wool, he suggested that the herdsman breed out the unfit and undesirable characteristics found in the city by pairing citizen-sheep of complementary temperaments (like the woof and warp of a piece of cloth) towards the production of a generation of obedient and agreeable offspring.[21] While the whole idea that we give politicians this kind of control over our lives may seem abhorrent, Plato saw it as a necessary step towards the production of an orderly and efficient *polis*. And, as the dialogue closes, we are told that this "royal science" which is the "greatest of all sciences" will produce a just city and a happy citizenry.

Two Cities

All told, the ancient Greek political philosophies of Plato and Aristotle present a rivalry between two visions of political rule: *Vision 1*, a kingly *techne* where the city and citizenry are treated like malleable material to be formed and controlled towards predictable ends in the same way a craftsman makes a product; or *Vision 2*, a *phronetic* rule where politics is guided by an overarching understanding of what makes for a good and happy life. It might be said that today we still face this same choice. If our goal as human beings and citizens is security in the predictable and controllable, then we should embrace the rule of the *technites*. Indeed, our current drive to understand and manage our bodies and minds through the "crafts" of neuroscience and genetic engineering suggests that we are not willing to leave anything to chance as we seek a thorough technical control of every aspect of our physical, emotional, and psychological selves. As things stand, we seem to have chosen the leadership of the craftsman-king.

Of course, it is not easy to know whether Plato was actually advocating for kingly *techne* in *The Statesman*, advising against it, or doing something else entirely. He even admitted that despite the fact it may be for the best, it is also highly unlikely that anyone would actually willingly submit to this sort of rule. So, even though there is something terrifyingly attractive about the efficiency of the craftsman-king's skill at expertly applying a dependable set of procedures towards the production of an excellent city or a happy life, the realization that the citizenry must be either fooled or coerced en masse into complying to his reign makes *The Statesman* more of a warning than an endorsement.

And while we would still not today voluntarily submit to a eugenics program that seeks the creation of a compliant citizenry, we do seem to accept the need for psychopharmalogical and genetic manipulation. In essence, we are allowing the *technites* to change our temperaments to conform to a certain standard of physical, emotional, and psychological health. Yet, is this not a very similar thing to what the *phronimos* does? The *phronetic* statesman does not simply facilitate the free realization of any of our goals or satisfaction of any of the passions, but necessarily ones he deems as "good." He seeks to influence us to conform to a given standard of physical health, emotional stability, and social behaviour, serving as a template for a good and proper way of living. It could be said that both the raw materials of a craftsman and a youthful citizen under the tutelage of a *phronimos* will not develop naturally or independently towards their end states (whether a piece of cloth or a good citizen/statesman). They both require something outside of themselves to move them in that direction.

Still, while there is an interesting similarity between these undertakings, there is also an important difference. To put it in Aristotelian terms, where a technician imposes an "external efficient cause," the *phronimos* encourages or activates an "internal efficient cause." To explain, where wood will never on its own accord shape itself into a chair, young men and women will, to some degree, change themselves willingly into good people provided that they are placed in the proper environment. In turn, where the craftsman imposes a form onto his materials, the *phronimos* can expect his "material" to change itself voluntarily to its new form.[22]

Incredibly, this parallels an important division Aristotle makes between things by nature and things by artifice. In the *Physics*, Aristotle explains that something is "by nature" only when it has in itself a source for change and staying unchanged.[23] In other words, its source of growth, movement or "efficient cause" is found somewhere within. For example, a tree grows from a seed to become a fully grown tree – the efficient cause of the tree is in itself. While it requires a proper environment for growth, it is still something within that seed (what we might now call DNA) that governs and compels the development of the tree. A pebble will never become an oak no matter the amount of sunlight, rain, and rich soil it is exposed to because it lacks that required internal efficient cause. A shoe or a house, on the other hand, has an "external" efficient cause in a cobbler or house-builder – in the maker and not in the thing made.[24] They do not "grow" or form without the imposition of the external agency of the craftsman.

We can conclude, quite reasonably, that trees are natural and shoes and houses are artificial. We can also conclude that things guided by *phronesis* are natural and those by *techne* are artificial. The *phronimos* is an essential part of the "proper environment" (the sun, rain, and soil) for the growth of good people, spurring the natural development of young men and women towards their becoming publicly engaged and responsible citizens. Differently, the *technites*-king attempts to produce or construct good citizens. If we were to submit to kingly *techne*, we would be placed in the same circumstance as the piece of wood waiting for the craftsman to form us into a chair or other product. As citizens of a techno-*polis*, we would never develop our own internal capacities for right, ethical, or lawful behaviour; never of our own accord will we be able to move towards the final goal of the good life. Just as a piece of wood will never by its own volition start to be more like a chair, we will never by ourselves become good people. Instead, we will require the ubiquitous presence of a hands-on *technites* forming and shaping us to be happy and satisfied. Like a chronically ill patient, the citizens of the technically derived city will need the external care of a doctor to ensure their health and continual development and well-being.

If we now reconsider the objection to *phronesis* that it is undemocratic, it seems that it at least facilitates the voluntary and willing compliance of the citizenry. *Techne*, by contrast, is necessarily domineering in its demands on the city and the citizen. In the chapters to come, we will see a movement away from the kind of politics that facilitates the internal mastery of the passions or activating the internal efficient cause of citizens to the external mastery of the passions through various technical means.

As we will also see, this movement from *phronetic* to technical rule introduces a new and disturbing problem. Aristotle explains that making (*poiesis*) and acting (*praxis*) are different, "For while making has an end other than itself, action cannot; for good action itself is its end."[25] Where the end of *techne* can "live on" long after it is made by a craftsman, as a horseshoe exists without the presence of a blacksmith, the activity stemming from *phronesis* is always directly linked to the *phronimos*. Where the good action of the *phronimos* cannot exist outside of the presence of a human being, the products of technical knowledge are independent from human life. In turn, technical knowledge can produce artefacts that have negative impacts on human beings, including the maker, and still count as the products of *techne*. We can and have made many products that are clear dangers to life: chemicals, weapons, and other technologies. This raises the possibility that the kind of community produced by technical rule could be dangerous or, at the very least, would not necessarily have to be good. In a different way, because it is always linked to the good behaviour of a human being, *phronesis* must always indicate action conducive to human life. From here, we could say that there is no impetus on a technical leader to make a city conducive to the good life. *Phronetic* leadership, however, is by definition directed towards the good life.

But, while this may be true of ordinary *techne*, it may not be true of kingly *techne*. Normally, once the craftsman is finished making a product, the properties of the materials used to make it are once again subject to to the forces of nature; matter restarts its natural movement. Once built, the wood of the chair will still rot if left in the same environment as a fallen tree on the forest floor. In other words, the technician does not normally impose a permanent form onto nature. However, in the context of a kingly *techne*, the craftsman remains a part of the product he makes; he is a part of the *polis*. So, where he is normally an "external" cause, in this example he becomes an internal cause. Conceivably, this would allow the techno-*polis* to "grow" in a way similar to a thing by nature, with the ruling craftsman constantly adapting and repairing his product to contingencies and new circumstances in the same way a tree adapts to new growing conditions. The *polis* of the craftsman-king takes on something of an artificial life, under the permanent imposition of an adaptive *techne*, never allowing any of its human material to return to their natural state.

Surprisingly, despite his advocacy of *phronesis*, Aristotle himself considers the possibilities of this kind of city. At one point in the *Politics* he even notes that "the primary factor necessary, in the equipment (*choregia*) of a state, is the human material."[26] This decidedly technical tone is accentuated all the more by the use of the word *choregia* to describe the "human material." Translated as equipment in this passage, *choregia* originally referred to a contribution of supplies or costumes given to the chorus of a play by a wealthy citizen. Here it implies that the people of a city are basic material for the city's production, no different than the wood and nails used by a carpenter.

However, it is important to note that Aristotle uses this term in his discussion of an ideal or imaginary city built from scratch. In this situation, we might consider the population an inert material that could be selected based on certain criteria and moulded at will. Similarly to Plato's *Statesman*, Aristotle advises that the "human material" of this ideal *polis* should be selected within a certain size (i.e., population) and be composed of persons of a certain natural endowment or temperament (e.g., intelligence and courage). As presented here, the maker of the *polis* can simply choose the amount and kind of people he wants in his city. But Aristotle, being of a practical sort, seems to put this consideration of the "ideal city" to a quick end, deciding in book 7 of the *Politics* that "it is easy enough to theorize about such matters: it is far less easy to realize one's theories. We talk about them in terms of our wants; what actually happens depends upon chance."[27] So, while it is good to think about what is ideal, politics is in many ways governed by what is given to us already.[28] Under normal conditions, no political leader can *simply* choose the citizens he is going to lead.

Still, the very fact that he conceives of the ideal city in such technical terms bears further consideration. For Aristotle, the "ideal" is not necessarily something that is out of reach or purely theoretical, but instead consists in having all the material conditions of life met as one would wish.[29] Under these conditions, it might actually be possible for a statesman to be like a master craftsman who imposes a form onto human material in the same way a carpenter imposes a form onto wood. Just as the carpenter understands how to turn wood into a chair, the legislator moulds the natural character of humans into good citizens.[30] And, even if fortune gives the statesman these materials "pre-prepared" (with their endowment and temperaments already set), this is really no different than the situation of any craftsman. Aristotle writes: "The art of the statesman does not produce human stock, but counts on its being supplied by nature and proceeds to use her supply ... It is not the business of the art of weaving to produce wool, but to use it."[31] Just as the weaver works with the already given nature of wool and the carpenter with the given nature of wood, the statesman must work with the given nature of the citizenry.

But, there remains a problem with this city. Aristotle also recognizes that "human material" behaves quite differently from the materials of cloth or a chair because it demands a say in the product it is a part of: the city. Imagine if the lumber of a house started instructing the house-builder on how to build a frame or the costumes of a chorus began to make suggestions to the actors. As we know, humans have many opinions on what is good and right. And, without doubt, this makes the building of a city far more complicated than the construction of a house, chair, or blanket. Instead of a craftsman with absolute control, a city requires a leader able to consolidate and direct the plurality of opinions, instructions, and suggestions that exist in the city's population. Without tolerance and flexibility, life in the city would be rigid and unable to adapt to the unexpected or unforeseen, nor would it allow for community-wide interpretation, discussion, and debate of laws.[32] In this light, the prospect of having a *techne* of human material or kingly *techne* is unlikely.

And yet all of this suggests that, if humans *could* be treated as are the materials of a weaver or a carpenter, the city could be run on precise technical knowledge alone. Expertise over "human material" would require a total effort to anticipate and control the thoughts and actions of every human. While this may be impractical or undesirable, it may also be that Plato and Aristotle both agreed that it was not impossible. Ultimately, though, due to the complexity involved, the singularly technical ruler, such as Plato's weaver of temperaments, is an impractical creature. Rather than his technical knowledge, the activity of the good statesman is exhibited by way of good judgment or *phronesis*.

3   The Decline of Good Judgment

This chapter explores medieval Christian conceptions of good judgment as they compare to Aristotelian *phronesis*. The objective here is to show how prudence becomes delinked from practical decision making and in turn is weakened in its role of limiting the influence of technical knowledge. Rather than a "bottom-up" virtue founded upon experiences of bodily pleasure, family, and community interaction, judgment is reconceived as a "top-down" virtue intended to order the soul, life in the city, and our relationship with the natural world.

While Aristotle tells us that *phronesis* is an intellectual virtue, it is most notable for its practical character. It allows us to properly navigate the myriad difficult choices we face every day. Rather than simply an obscure concept hidden in the pages of a philosophical text, *phronesis* is on clear display in the successful lives of friends, neighbours, and colleagues. People with this virtue are not simply lucky or blessed to live a good life, but have deliberately, knowingly, and meritoriously chosen a path that leads them to happiness. They are not simply clever, able to fool people into believing they are upstanding citizens. Instead, they are motivated to act and live in a virtuous manner. They can do this because they have a general understanding of what is good and can apply that understanding to the complex and particular conditions of everyday life. So, *phronesis* requires at least some access to or ability to grasp a universal concept of "the good."

While surprisingly difficult to explain in philosophical terms, in practical terms "the good" is a straightforward thing. Anything that results in happiness, flourishing, prosperity, pleasure, in other words what amounts to living "the good life," can be understood as a realization of "the good" in this world. So, when it comes to Aristotle's description of *phronesis*, we can put aside the modern criticism of the classical philosophical tradition mentioned in chapter 1 that suggested the classical concept of "the good" is indemonstrable, a purely abstract notion with no tangible qualities to evaluate or agree upon. Far from it, the good is obvious in the lives lived by our most virtuous citizens.

For Aristotle, being virtuous was not simply about possessing a certain set of personality traits but was something that was done or acted upon. It was by definition excellent action for all to see. A soldier was courageous because of his actions on the battlefield. A citizen was generous because of the amount of money he donated to festivals and civic works. And a politician had good judgment because of the effectiveness of his legislation. By this notion, there was no such thing as "inner beauty" for the Greeks. Beauty was something out in the open. From here, we can see why Aristotle thought that the best social norms, ethics, and laws of the *polis* must be drawn from the experience of living together and deciding together about what was good and what made citizens happy. This simple idea of virtue as something that appears to or is recognized by fellow citizens will come up later in chapter 7 in relation to the possibility of teaching and learning virtue today.

Of course, just because Aristotle understood virtue to be something seen does not mean that there was to be easy agreement on what any of these virtues actually were. In fact, it was the very difficulty of deciding what was virtuous or what counted as an ethical action that explains why he thought that the practice of politics was so critical for the Greeks. Politics was not something that occurred distantly in the legislature or during periodic elections, but was deeply interwoven into daily life. Politics was the way to come to an agreement on what was virtuous and how to create laws, institutions, and educational curricula to help support the practice of these virtues in the community. This is why, for example, he viewed tyrannies as deviant or unnatural political regimes. Without a forum for public discussion and judgment, without politics, there was simply no way for individual citizens to contribute to an understanding of the common good. Necessarily, the practice of the virtues endorsed by the tyrant would be unlikely to satisfy the desires and aspirations of the citizenry. And *phronesis* is the key to all of this. It is the capacity to re-evaluate ideas agreed upon previously about virtue and adapt them to changing circumstances on the ground, compelling the community to modify their norms, institutions, and laws accordingly. For Aristotle, a *polis* paralysed by convention and tradition would inevitably fall short of the good life and make its citizens unhappy.[1]

But, we may ask, what if there can be no good life after all? What if we have no access to the good and thus have no capacity to apply it to our everyday lives? Looking at the misery, poverty, violence, war, and death that plague the world, it might seem that the good is an illusion or, at the very least, an otherworldly concept. With this in mind, many people might feel that the whole idea of *phronesis* represents a kind of overconfidence, arrogance, or dangerous elitism – the notion that someone can so successfully determine the course of their own lives while being both decent and happy is hard to believe. We are simply too selfish, greedy, and unethical at heart to allow for anything like the virtuous and incorruptible *phronimos* to exist in our society. Too often we learn that those that we thought were the greatest and most noble among us, the seemingly magnanimous, turn out to be just as disappointingly fallible, weak, and fickle as everyone else. It may in fact be that the *phronimos* of old is merely a creature of philosophical imagination after all, rather than someone who actually walks among us.

While we may lament this pessimism and distrust of human goodness, we can at least take some solace that it is not simply a function of our cynical contemporary society. In fact, the degradation of practical reason has quite a long history. The Judaeo-Christian tradition, the other great pillar of Western civilization that stands beside the ancient Greek, never shared Aristotle's high regard for human decision making. This is not to say that the monotheistic religions of the Old and New Testaments have no concept of the good life, only that humans in this world could never wholly achieve it in the here and now. Early Christian thinkers, for example, articulate a profound lack of confidence in the ability of human beings to *by themselves* judge and act towards what is good and right. Alternatively, they describe good action as accepting God's grace or selecting a pre-existing set of divine instructions imprinted on all of us.

St Augustine: Two Prudences

This dismissal of our practical capacity for good decision making was one of the ideas put forward by the great scholar and Christian philosopher St Augustine (AD 354–430). His less-than-enthusiastic view of human accomplishments was somewhat understandable considering that he lived through the sacking of Rome by the Visigoths in 410. The realization that the "Eternal City," supposedly the very manifestation of heaven on earth, could be overrun and looted by barbarians shook the Roman Empire to its very foundations. Augustine, however, took this event as a good reminder of the true nature of Christianity and spent the rest of his life attempting to reassure his fellow Christians that the political and material success of Rome was not at all the point of their religion, that they should go on believing even though their livelihoods and possessions had been destroyed and stolen. If anything, he thought that everyone should thank the pagan plunderers because the corporeal delights so easily acquired in a cosmopolitan city were a distraction from living a proper life in line with Christian values.

In truth, Augustine had himself come to this conclusion well before the sacking of Rome. He recounted in his *Confessions*, which is thought to be one of the first autobiographies written in the West, that as a child his soul was in "ruins" and as a youth his body was wracked by "carnal corruptions" that led him to crime and to lust. He goes on to say that, in 387, at the age of thirty-three, he renounced the pleasures of the flesh that surrounded him everywhere, was baptized by his mentor Ambrose of Milan, and finally left the sex and material gratification of his younger years behind him. Augustine realized that his worldly ambitions, whether for career success, beautiful women, or material things, would never provide him with true happiness; and that the restlessness of his life and the disorder he felt in his soul could only be remedied by turning away from the earthly things so valued by Roman society. Describing the moment when he is tempted one final time by possessions, vanities, and his old mistresses, he asked himself, "Do you think you can live without them?"[2] With his answer of "yes" he pushed aside the whole concept of virtue as it was practised by the Greeks and adopted by the Romans, making room for a new way of living that would dominate the Western world for centuries to come.

Following Augustine's lead, Christians utterly reject the idea of "the good" as it was expressed by the Greeks. Both theologically and practically, they argued that the world they lived in was far too wicked to deliver anything like the good life. The fall of a hedonistic, debauched, and violent Rome was a great illustration of this fact. Of course, with this rejection the Christians also rejected the old pagan idea of *phronesis*. In a world thought to be full of overwhelming corruption and temptation, it is no longer thought possible for an individual to be so capable of judging the right thing to do. For Augustine, we are faced with only one critical and constant decision. Every day we must decide whether to choose the ephemeral satisfaction afforded by pleasure and sin or to accept God's grace into our lives in pursuit of a deeper happiness. In other words, rather than the easy choice to satisfy our bodies, we must instead make the much harder choice to gratify our souls.

To emphasize this important distinction between bodily and spiritual satisfaction, Augustine argued that there are two distinct and opposing kinds of judgment. In his most important work, *City of God*, he describes the difference between "prudence of the flesh" and "prudence of the spirit." As its name implies, the first prudence lets us satisfy our carnal wants and desires, leading us down the path of sin. The second prudence allows us to live a proper, religious life in harmony with God. All told, Augustine reminded his readers that the tremendous inventiveness and considerable effort that goes into rewarding the flesh was a distraction and should be redirected towards the ascetic pursuit of living a spiritual life. Perhaps he explained this most succinctly in a frank letter he writes to the archbishop of Arles, Honoratus, in 412:

> There is a certain life of man involved in the carnal senses, given up to carnal joys, avoiding carnal hurt, seeking carnal pleasure. The happiness of this life is temporal: to begin with this life is a matter of necessity; to continue in it a matter of choice. Doubtless the infant issues forth into this life from the womb of its mother; as far as it can it avoids the hurts and seeks the pleasures of this life; nothing else counts. But after it reaches the age at which the use of reason awakens and its will is divinely aided, it can choose another life whose joy is in the mind, whose happiness is interior and eternal. Truly there is in man a rational soul, but it makes a difference which way he turns the use of reason by his will: whether to the goods of his external and lower nature, or to the goods of his interior and higher nature; that is, whether his enjoyment is corporeal and temporal, or divine and eternal. This soul is placed in a middle state, having below it the physical creation and above it the Creator of itself and its body.[3]

So, instead of lower goods that provide outward carnal pleasure, the purposeful turning of the will to goods of a spiritual nature will instead deliver us an "interior and eternal" happiness. Needless to say, the identification of these two separate and opposing choices or "prudences" stands in stark contrast to the relationship Aristotle describes between making decisions about satisfying the lower appetites and making more important ethical choices later in life. Rather than putting us off or distracting us from a pursuit of higher things, Aristotle thought that learning how to properly satiate our bodily needs, learning moderation, for example, laid the groundwork for a life filled with good decisions. But for Augustine, rather than directing or guiding lower prudence to decisions about higher things, adherence to prudence of the spirit instead attempts to free us from the immature, irrational, and distracting influence of the prudence of the flesh. The passions, appetites, and emotions that Aristotle thought served as a foundation for later good decision making are now considered conduits to dangerous temptation and transgression. Now, the pleasures we derive from sex and food are no longer an indication that we are living a good life but the very opposite: a life of sin.

Prudence of the spirit allows us to direct our actions away from the evil in the world and within us towards that which is good. Where Aristotle believed human fulfilment came from navigating the complexities of everyday life towards virtue and happiness, the early Christian thought of Augustine presented excellent action and corresponding human goods embodied in a life of spiritual harmony with the Christian God. Ultimately, it is God's love, grace, or divine intervention that guides us to choose the right way and avoid that which leads us astray to the things of the flesh. As Augustine says, it is from Christ that we learn what we are to love and how much we are to love it and that God is our highest good.[4] For him, wise choices are no longer a matter of finding the right way to handle the complications of one's particular circumstances, but instead making the choice to embrace a predetermined, divinely written course of action. Consequently, the unique position of practical judgment as a bridge between lower goods and the higher goods is broken. Yes, Augustine was not so miserable as to say that we must or even should avoid all bodily pleasure lest we suffer eternal damnation. The point, however, is that this pleasure does not inform or assist us in choosing the way to God. He is asking us to turn away from the petty and practical things of the material world towards the more truly satisfying life of the spirit.

Critically, this splitting of prudence also opens up a big divide between our external life in the city and the "interior happiness" Augustine writes of in his letter to Honoratus. Instead of seeking fulfilment in the here and now, we are asked to look upward with a faith that there is a perfect "City of God" waiting for us once we pass from the misery of this profane and corrupt "City of Man." Augustine is often described as a Neoplatonist because, similarly to Plato's discussion of the metaphysical good, he posits the existence of a perfect supranatural standard or prototype for a perfect world. However, there seems that there is nothing we can do – no deeds, no politics, and no philosophy – that will help make our corrupt world any more like that perfect world.

The division between worldly actions and Christian morality leaves ethical and political education in a tough spot. Unlike the *phronimos*, who is perceptibly rewarded for wise ethical and political choices, the good Christian may have quite a different lifestyle: a poor and dreary existence, subordinating physical desires to the noble and intellectual pursuit of knowledge of God. And, while Christians may be profoundly happy in this effort, there also may not be much in the way of public evidence for their happiness. While Aristotle's *phronimos* would necessarily have the respect and admiration of his colleagues and neighbours, Augustine's good Christian might just as likely live a lonely existence with no material prosperity or other bodily satisfactions. Where the *phronimos* is an ethical exemplar and a political leader whose actions and choices are to be imitated by young citizens, the model provided by the good Christian may be harder to follow. Because there is no obvious or outward display to imitate, and because there is no worldly reward for ethical decisions, the student of Christianity can only come to the choice to follow God by faith.

Tellingly, just before his decision to convert, Augustine drew great inspiration from his readings about the life of one of the very first monks, Anthony of the Desert, who lived alone in the wastelands of western Egypt around the turn of the third century AD. The central idea behind the monasticism of Anthony was to remove oneself from the temptations of society and worldly affairs to focus solely on spiritual pursuits. For many Christians, Augustine among them, the ascetic and lonely existence of the monk was what replaced the pleasure-seeking and public life of the *phronimos* as the ideal to which all humans should aspire. What a difference a few centuries make.

The celebrated life of St Paul the Hermit (died c. 341), who died a few years before St Augustine was born, is another example of the Christian rejection of the old ideal of the good life. His story is recounted in St Jerome's *Vitae Patrum* or *Lives of the Fathers*. By way of introduction to the life of Paul, St Jerome describes a time when Christians were being cruelly persecuted at the hands of the pre-Christian Roman Empire. The Romans, seeking complete loyalty to the emperor, went to great efforts to break the spirit of the growing members of this defiant religious sect. Here, St Jerome describes one outrageously brutal attempt:

There was one particular martyr who persevered victoriously in the faith through tortures by racks and hot metal plates, so they ordered him to be smeared all over with honey and laid him out in the heat of the sun with his hands tied behind his back, hoping that even though he had survived the hot frying pan he might succumb to the burning pain of the insect bites.[5]

Taking a different tack, the Romans tied another unlucky fellow to a bed and tempted him with a prostitute. But he refused to give in to sinful pleasure and, according to St Jerome, instead "inspired by heaven, he bit off his tongue and spat it in her face as she tried to kiss him ... So the immense pain which followed was stronger than the feeling of lust."[6] In either case, the Romans were frustrated in their efforts because both of these men were able to muster the strength to turn away from the corrupting things of the flesh and become impervious to the physical torturing of their bodies, whether via intense pain or pleasure. As a young man, Paul himself sought to escape similar persecution and fled to a cave in the mountains, where he lived a more or less solitary existence of a hermit for the next ninety years, eating nothing more than bread and drinking nothing more than water, dying at the ripe old age of one hundred and thirteen.

For Jerome, the point of these stories is that the lives of these men, especially that of the monastic Paul, are examples of Christian ethics in practice, demonstrating that the good life is not based on worldly standards but rather on a life of sacrifice and meditation. Where it may seem to the non-believer that Paul was living anything but the good life, alone and in extreme poverty, he is nonetheless taken to be a religious exemplar for all Christians. If nothing else, Jerome's great praise for the life of Paul again highlights the considerable difference between the ancient Greek and Christian traditions. As Alasdair MacIntyre puts it in *After Virtue*, "For the New Testament not only praises virtues of which Aristotle knows nothing – faith, hope and love – and says nothing about virtues such as *phronesis* ... it praises at least one quality of virtue which Aristotle seems to count as one of the vices relative to magnanimity, namely humility."[7] That is to say, whereas the Greeks praised the generosity of the wealthy,[8] the early Christians see "the rich destined for the pains of Hell." The point is not so much that wealth and prosperity are sins, but rather that they are manifestations of a life spent focused on worldly or material possessions as opposed to meditating on God.[9]

For these Christians, ethics are not derived from a life lived in a community or city, but rather from individual meditation on theological principles. But because almost no one has the advantage of decades of solitary reflection, it is nearly impossible to follow the extreme example of Paul. Instead, they have to deal with their families, neighbours, and colleagues. Here, we are left with an outstanding question similar to one asked by the Greeks. How can we agree on a *common* expression of virtues when virtue is itself based on a tremendously demanding and profoundly individualistic faith in an inscrutable God? Because early Christian interpretations fail to understand good judgment as public action, and instead contain it within the realm of individual spirituality, the ethical actions of citizens in the city are separated from their theoretical understanding of what is good and what truly makes them happy. In turn, politics can no longer be an institution that directs citizens towards the good life. At its best, politics should facilitate the journey of each individual towards choosing happiness with God or, at the very least, get out of the way.

St Thomas Aquinas: The Spark of Conscience

At this point we might again be reminded of the modern criticism of religion: the supranatural or external standard of God is indemonstrable and thus cannot serve as a reliable or common foundation for ethics, law, or politics. Because early Christians seem to dismiss any and all corporeal manifestations of the "good life," there may be no way to verify or agree upon its quality or content. The best we can do is to believe that we are living a good life rather than proving it; the truth of one's goodness will only be revealed in the next world. (In fact, Augustine even warns against taking motivation from the promise of heavenly reward, arguing it is selfish and prideful to assume to understand God's will!) For Augustine, ascetic prudence of the spirit is the only allowable route to the good life, a single choice between eternal salvation and avoiding eternal damnation. In turn, all the contrivances of the material world, including family, community, and politics, are to be put aside, ultimately to be sacrificed for this highest of ends. Only by fully accepting God's unrequited love can one travel on the path to the good life.[10]

So, even though there is a host of broad Christian values derived from sacred doctrine that are suppose to help individuals make good decisions in their daily lives, there may be little in the way of specific instructions on how to apply those values to one's own particular life and community. Even today church leaders and politicians endlessly debate how to properly apply religious values as a way to validate or condemn certain lifestyles and laws.

Centuries after Augustine, medieval Christian thinkers were still trying to solve the dilemma of how to live a moral life in conformity with theology. The leading philosopher of the day, St Thomas Aquinas (1225–74) was fortunate enough to be able to turn to the work of Aristotle, which had recently been rediscovered in the West and translated into Latin. In fact, the dominant theological school of the time, scholasticism, was almost entirely focused on reconciling Aristotelian philosophy with Christian theology. With Aristotle's *Ethics* and *Politics* firmly in hand, Aquinas might have concluded that the source of the problem lay in Augustine's splitting up of good judgment into two separate realms that left our decisions about how to satisfy our worldly appetites disconnected from decisions about how to live a good and ethical life. The solution then might be found in its reunification. Yet, rather than reunifying, reviving, or repairing the fractured Aristotelian virtue of *phronesis*, Aquinas sees an answer in two new virtues: *prudentia* and *synderesis*.

Most often, the Greek word *phronesis* is translated as prudence, itself taken from the Latin word *prudentia*. However, there is an important difference between the Greek term used by Aristotle and the Latin term used by Aquinas. As one scholar reviewing medieval conceptions of prudence notes, "Prudence, as distinct from *phronesis*, becomes a virtue of the inferior part of the soul, whose main function is the habituation of the rational faculties in the governance of the passions."[11] Rather than describing the ability to render a general understanding of the good into a specific practice, as is the case with Aristotelian *phronesis*, Aquinian prudence is about good habits. Broadly taken, we are prudent when we exercise good habits in the satisfaction of the appetites, in the management of our households, and in the practice of politics. We learn how to do these things from trial and error and the lessons of our parents, teachers, and political leaders. This prudence allows us to do simple things like choose a healthy side salad over French fries, or exercise over the couch, or fidelity over infidelity, and so on. Interestingly, according to Aquinas we do not make these choices for any direct personal, familial, or communal reward but simply because we reason that they are *the right thing to do*. In other words, these choices do not have to result in pleasure of any kind, and in some cases may deliver quite the opposite: pain and discomfort.

Problematically, though, no matter how many good habits we think we might have, it may be hard to actually know if one is in fact *doing the right thing*. What if our reasoning is without any merit? Or deviant? Or the product of some delusion? What if we had bad parents who taught us bad habits? What if our understanding of something like bodily health is based on faulty information (think of all the conflicting evidence about what constitutes a healthy diet!)? Indeed, even a serial killer may truly think that what he is doing is the right thing to do. Again, because there is no definite or expected outward reward for "good" habits, no necessary sensual pleasure or payment, we cannot actually know for sure that we have made truly good decisions.

In an effort to address this difficulty, Aquinas explains that we possess a higher faculty by which we can be reassured that this "governance of the passions" is in fact directed towards something good: *synderesis*. While medieval scholars explain it in various ways, typically *synderesis* suggests something like a "spark of conscience" that tells us whether our actions are good or evil. Aquinas himself describes it as the imprint of God on every human soul, explaining that our "reason is instructed by the Holy Ghost about what we have to do."[12]

Therefore, while individuals can still act in an imprudent manner, choosing through their free will to turn away from what they know is right, they can never simply dismiss *synderesis* – like an angel sitting on their shoulder, the spark of conscience is always with them whether or not they choose to ignore it. So, even though the French fries may make us feel good at the time, we all the while know that despite our great pleasure we will still feel guilty for having eaten them, all thanks to *synderesis*. The point here is not that we should deny ourselves all pleasures, but that we should subordinate our desire for things like tasty and fatty foods to the more important desire of maintaining good health. This divine imprint on our souls informs our conscience and compels us to act in a prudent manner. As Aquinas writes, "*Synderesis* moves *prudentia*."[13] So, where *phronesis* might be called a "bottom-up" virtue acquired through experience, Aquinian prudence is a "top-down" virtue taking its cues from the superior *synderesis*.[14] While we can still learn particular habits from our friends, parents, and teachers, we only know they are good habits because of our conscience. And, if we do choose to ignore our conscience and give priority to bad habits and sensual pleasure, then we are sinful, exercising a deviant "*prudentia* of the flesh" that is free of the guiding influence of *synderesis*.

This might give the impression that this distinctively Christian virtue is like a direct line to the voice of God giving us instructions on how to behave. Unfortunately, it is nowhere near that simple. Aquinas explains that the proper ordering of the desires as well as the general principles of a moral life are informed by "natural laws" permeating all of God's creation. Natural law was itself not a new idea. The ancient Greeks distinguished between the laws of nature (*physis*) and the laws of men (*nomos*). Similarly, the Romans adopted a concept of natural law (*jus natural*) that was shared by all living creatures distinct from the "law of nations" (*jus gentium*) of states.[15] In both of these earlier conceptions, there is an effort to work out which parts of human behaviour belong to one realm or the other. Aquinas's discussion of natural law, by contrast, is part of his larger description of the grand design of a universe ordered in a great chain of Being. In surprising detail, he attempts to explain how everything and everyone falls somewhere in this hierarchy, with the higher positions having a natural authority over the lower. It was not only an attempt to explain why humans have dominion over animals, but also a justification for the Church's power over the masses and monarchs' divine right to rule over their subjects. In Aristotelian-like terms, Aquinas concludes that when our lives and communities are properly ordered, when everyone finds their proper place within this hierarchy, humanity will be living in accordance to God's plan and in harmony with natural law.

Unfortunately again, medieval theologians also concede that the exact content of these laws remains a mystery. Unlike animals that can instinctively find their position within the natural order, humans are left to figure it out for themselves, employing our unique capacity for reason to discover our proper place in creation.[16] For Aquinas in particular, the path of reason sets our minds and our souls away from the corrupting influences of bodily pleasure and towards acquiring this most precious and sought-after piece of knowledge. In fact, this quest for knowledge undergirds almost all of the intellectual enterprises of the period. As a result, in medieval universities across Europe, theology students begin to comb through scripture for secrets buried within its passages. By pulling out these hidden meanings, it was thought that they could actually come to understand existence.

Furthermore, because the whole idea of natural law implies that God's plan is written everywhere and into everything, there are also efforts to discover it in the very make-up of the physical world. So, astronomers probe the skies intently to find any sign of a divine design in the arrangement of the stars. Alchemists begin experimenting on various substances with the hope that they could release hidden truths locked within matter. The Dutch alchemist Johannes Isaac Hollandus explained the purpose of his experiments in his study *Opera vegetabilia* (ca. fifteenth century): "Now I will teach and describe the secret of the arts, which secret is at the heart of all secrets hidden in the art of alchemy; since one will here understand the wonderful works that God has accomplished in all things he has made."

All told, the spiritual mission to know God also ends up spurring a tremendous effort to uncover the mysterious laws of nature. While more of a religious quest than a consequence of intellectual inquiry, medieval science sets out to find the true order of the physical world and prove that there is a supreme intelligence behind it all.

Critically, this same idea also led to an effort to try to find the natural laws that govern human beings. Along with seeking out the reasons behind our most basic instincts such as the innate desire for self-preservation and the need of men and women to have and raise children, there was a concerted effort to explain how humans are supposed to live and interact in society. It was thought that this discovery could inform the making of "positive" laws and institutions that would help human beings better conform to the natural order. Aquinas, for one, explains that this conformity requires the coercive power of government because, in their state of original sin, human beings have lost their instinct to obey authority.

Finally, natural law also provides a rather precise model for individuals to follow. It is here that the virtue of *synderesis* really comes into play. It is what enables us to recognize and understand practical moral principles. It is what allows us to know that acts of charity, love, and courage are good when we see or experience them. Again, putting this understanding into action does not come easily. Our tendency to sin, the endless opportunities for corruption, and our fallen nature make living a moral life a tremendously difficult task. And so we need the secondary virtue of *prudentia* to figure out how to articulate these discovered principles in the changing and particular circumstances of our everyday lives, all the while keeping our wickedness in check. Importantly, the goal is not individual self-expression, but instead a conformity to an already existent template of the good life. In the *Summa*, Aquinas writes: "The right ends of human life are fixed;"[17] that prudence has its source in an understanding of those ends;[18] and that "prudence, which denotes rectitude of reason, is chiefly perfected and helped through being ruled and moved by the Holy Ghost."[19] What Aquinas is really describing then is the existence of a divine blueprint for the perfect life composed of fixed and universal standards for living.

All told, Aquinas sets up three interrelated projects. First, we embark on a new project of natural science to discover the hidden properties of the physical world. Second, we begin the project of a new political science to find the proper way to govern. Third, we engage in a project of moral science to correct our own sinfulness. A certain antagonism is brought to all three projects because in each case the world as it is given seems to be working against us, keeping something from us, and trying to distract us from truth and happiness. This antagonism is perhaps the most critical difference between Aquinas and Aristotle and more generally the Christians and the Greeks. Whereas Aristotle thought that the laws that govern nature and man could be understood through observation, discussion, and contemplation, Aquinas asks us to dig for the truth buried and hidden somewhere deep within creation.

In all of this, we see the beginnings of an effort to quantify and control the world and ourselves. By studying and manipulating external and internal nature as they are given, we attempt to overcome their corrupting influence and change the composition of nature to better conform it to the divine blueprint. The human capacity to change the material, political, and moral conditions under which we live also shifts human judgment onto new ground. Judgment as *prudentia* becomes the handmaiden of an already existent but hidden set of fixed laws. Now, the goal of judgment seems to be to bring these hidden laws out into the open, to change the fallen world of men to conform to their higher standards.

And yet we can only take this idea so far. For Aquinas, we cannot actually create the conditions of "the good life" in this world. True fulfilment is found elsewhere. Rather than being found in wealth, sex, power, honour, or fame,[20] Aquinas claims that perfect happiness resides in "the vision of the Divine Essence"[21] seen only after we die. Where Aristotle writes of *eudaimonia*, a life of well-being through the practice of human goods, Aquinas describes the ultimate end of humans as *beatitudo*, blessedness or union with God. Where in the Greek conception the good life is on clear display in this world, the medieval Christian conception saves it for the next. And, while Aquinas concedes that a certain imperfect happiness can be had in the here and now through study and work, ultimately "perfect and true Happiness cannot be had in this life."[22] In a similar way to Aristotle, who describes the experience of happiness in relation to the fulfilment of our *telos*, Aquinas sees a semblance of happiness resulting from our moving closer to God through good acts and thoughts aided by the virtue of prudence. But, *prudentia* will always fall short in some way, never really being able to fully articulate into action the plan God has imprinted on our souls. There is no equivalent to *eudaimonia* in Aquinas, because no matter how good we try to be we simply cannot exclude every evil or temptation from our daily lives.

As it turns out, *prudentia* in service to *synderesis* ends up working in a rather negative way. Unlike *phronesis*, it has more to do with clearing our souls of the infinite temptations and vices that obstruct or hinder our route to happiness, or as Aquinas says, "the removal of obstacles to the movement of love towards God."[23] It calls our attention away from the things of the body, the family, friends, and the city itself, clearing a path to the ultimate good found in a beatific vision of the divine. So, what was supposed to be a practical virtue that helps us navigate everyday experiences instead sets us on a rather impractical if not impossible course.

It is with this in mind that Aquinas decides that "prudence of the flesh ... is a sin, because it involves a disorder in man with respect to his last end, which does not consist in the goods of the body";[24] that "without the body the soul can be happy";[25] and even that "the fellowship of friends is not essential to Happiness."[26] Necessarily, the responsibilities of not just friends but also of parents, neighbours, teachers, community, and political leaders in the proper upbringing of young citizens are downgraded, subordinated to the invisible but ever-present instruction of the Holy Ghost. While we may not like the idea of a reduced role for families and communities in the development of good people, the flipside of this emphasis on our individual relationship with God is that a source for ethical behaviour will remain available to everyone even in the worst of conditions. Even if a person has rotten parents or is stuck in a corrupt community, he or she can still choose to accept God's love into their lives. We might relate this to a similar idea at work in contemporary discussions of human rights. Regardless of the regime or custom under which a person lives, everyone is morally entitled to certain freedoms of action and thought. For the most part, this opportunity to escape or overcome the circumstances of one's birth and upbringing is not available in Aristotle. For him, a deviant and unhappy *polis* will necessarily result in a deviant and unhappy citizenry.

Now the path to good decisions has become a little less complicated. The human capacity to make ethical choices is no longer a "bottom-up" exercise dependent on the particularities of our bodies, families, and friends, or the tremendous variety of experiences of life in the city, but instead depends on the singular "top-down" virtue of *synderesis*. With no necessary *fundamental* ethical connection between the individual and her community, the "good life" is something we can all strive for (and nonetheless fall short of), regardless of circumstances on the ground. The fixed point in the heavens by which we can gauge our ethical compasses is uniformly available to all humans regardless of where, how, or with whom they live.

This said, it should come as no surprise that, despite his general agreement with Aristotle that a man living outside of the *polis* will lack the proper means for living, Aquinas does not actually believe that politics is an absolute requirement for human goodness. In his *Commentary on Aristotle's Politics*, Aquinas returns to the familiar figure of the friendless monk as the best example for humanity, calling St Anthony the Hermit a superior sort of human being because he had self-sufficiency "without human company."[27] But if Anthony, like Paul, is truly our exemplar, then *prudentia* has more to do with clearing our souls of the infinite temptations and vices that obstruct or hinder the route to goodness than with making good decisions about the city or its citizens. Like Augustine before him, Aquinas sees the unencumbered Hermit, without desire and alone, as the most prudent man. But, because the rest of us do not have the fortitude to live like this, we are left to contend with life in the city, doing our best to govern our passions, relate to our neighbours, and heed the demands of our political leaders. Again, in these dealings *prudentia* seems to help us only in a negative sense, removing obstacles rather than allowing us to see "what sorts of thing conduce to the good life in general," as is the role of *phronesis*. Where Aristotle sees the satisfaction of the appetites, time spent with friends, and the practice of politics as good ends in themselves as well as foundations for the teaching of ethics to the next generation of citizens, Aquinas sees them as subordinate to theology. At best they facilitate the journey to God and at worst they are great distractions. So, while still an important virtue for everyday living, prudence has substantially declined in stature.

4 The Rise of Technical Knowledge

The main point of this middle chapter is to consider how, after the de-linking of practical decision making from good judgment, technical knowledge comes to dominate our thinking and action. The "Great Reversal" of the judgmental and technical then follows from the new modern capacity to focus on human ends freed from the spiritual concerns that limited the practice of medieval Christian prudence. In a sense, the modernized prudence is a product of technical knowledge that allows us to begin the enterprise of ordering and controlling both the natural world and human nature.

Niccolò Machiavelli: The *Prudenza* Remedy

What we get from the Christian treatments of prudence is a strict division between the sacred and the profane and the spirit and the flesh. Mediating between these two realms, good judgment is put into service to control our sinful nature. The hermit is lionized because he is the most able to discipline his body, controlling his passions, appetites, and human nature itself. We could say that Aquinas understands politics as a communal effort towards the same, controlling and ordering the whole of the population towards their spiritual welfare.

Of course, controlling and ordering nature is also one of the key ideas of the modern era. But rather than being motivated by an effort to deliver divine moral principles into this world, modernity is characterized by the effort to satisfy decidedly temporal ends. This shift in focus from the otherworldly to the worldly was an element of a larger cultural and philosophical movement called "humanism." In part, humanism was prompted by a series of crises in the western Catholic Church about a century after Aquinas's death.[1] This led to an effort at change within the Church, including Martin Luther's provocative Ninety-five Theses (1517) that incited the Reformation and the eventual end of Catholic domination of Western Europe. But, these events also led many people to search for ethical guidance in places outside of religion. There was a new exploration of a *secular* humanism that challenged the Christian view of humans as sinners in need of salvation. Reason rather than faith, scientific investigation rather than solitary meditation, and a curiosity about reality rather than the divine became the central features of the effort to recover and revive the intellectual and cultural capacities of human beings that had been lost since the onset of the Dark Ages, which began centuries earlier with the sacking of Rome and St Augustine's call to turn away from the things of this world.

In fact, because of his emphasis on rationality, politics, and law, it is often argued that Aquinas was himself a humanist. While Aquinas did recognize a certain freedom in human affairs and choices, he viewed the goals and achievements of humanity as still hierarchically subordinate to God. The humanism that spurred the Renaissance, the Enlightenment, and modernity itself, however, went further. This change is well expressed by the infamous Italian political thinker and leading humanist Niccolò Machiavelli (1469–1527) when he observes in *The Prince* (1516):

> As I am well aware, many have believed and now believe human affairs so controlled by Fortune and by God that men with prudence cannot manage them – yes, more, that men have no recourse against the world's variations. Such believers therefore decide that they need not sweat over man's activities but can let Chance govern them. This belief has been the more firmly held in our times by reason of the great variations in affairs that we have seen in the past and now see every day beyond human prediction. Thinking of these variations, I myself now and then incline in some respects to their belief. Nonetheless, in order not to annul our free will, I judge it true that Fortune may be mistress to one half our actions but that even she leaves the other half, or almost, under our control.[2]

Considering his reputation as an arch-realist and the father of modern power politics, what is most surprising about this passage is Machiavelli's acceptance of the still large role played by God or "Fortune" in the actions of men. Rather than completely casting off the limitations placed on human achievement associated with medieval thought, Machiavelli instead presents a conflict between the goals of men and the authority of God. In this sense, he is a transitional figure straddling the medieval and modern eras. The transition, though, is momentous in that humanity is no longer subordinate to the divine but "almost" on an equal footing.

This conflict between the manoeuvrings of fortune, or *fortuna*, and the human practice of the virtue, or *virtù*, of prudence is an underlying theme of *The Prince*. Because Italy at the time was subject to such "great variations in affairs," including the invasion of foreign armies and shifting political leadership, it appeared to many Italians that the future direction of their communities was out of their control. It may have seemed that the spinning wheel of the goddess *Fortuna* had turned against them, leaving them helpless in the face of overwhelming circumstances "beyond human prediction." As a political treatise, *The Prince* instructs that this turmoil actually presents an opportunity for change; that fortune could be turned in the Italians' favour. A strong leader with the right skill and strategy could defeat the foreign invaders, establish political stability, and take control of Italy for the Italian people.

Taken this way, the concept of *fortuna* becomes more difficult to understand, indicating sometimes an unknowable, transcendent godlike force that is impervious to human influence and at other times merely a set of contingencies that with enough foresight and skill can be anticipated and neutralized. Whether *fortuna* is one or the other seems to depend on the practice of *virtù*. Instead of the Greek or Christian virtue, we have Machiavelli's novel *virtù*: not a set of principles, morals, or ethics but more simply the capacity to act towards a desired end, whatever it may be. For Machiavelli, if you are very virtuous then you will be able to achieve your goals most often, and chance, bad luck, and unfortunate circumstances will have less of a role in the direction of your life. If you have little virtue, it may seem that you have no say in the path your life takes and are helpless in the face of forces greater than you. So, there is no strict division between the realms of *fortuna* and *virtù*, but instead a constant battle for territory between them.[3]

Notably, the *virtù* under consideration in the passage above is prudence or *prudenza* in the original Italian. This secular definition of judgment, cut off from the guiding influence of either "the good" or God, owes its existence to the millennium-old splitting of prudence into two. Machiavellian *virtù* is Augustinian "prudence of the flesh" freed from its dour partner "prudence of the spirit." And, unlike Aquinian *pruden-tia*, *prudenza* is cut loose from the guiding influence of *synderesis*, freeing human action to work towards strictly human ends. Now, the tremendous energy put into material, political, and moral projects to uncover God's divine plans and conform the world to them is put into the service of the strictly human enterprise of quantifying and controlling the world, bending it to conform to the will of men.

So, while they might think they are powerless, Machiavelli is urging the Italian people to be prudent. The Romans, he reminds his readers, were successful and powerful because they were able to foresee and direct the flow of future events:

> In these instances the Romans did what all wise princes do: these take thought not merely for present discords but also for future ones, and the latter they forestall with every sort of ingenuity; when foreseen far ahead, discords easily can be remedied, but when you wait until they are upon you, the medicine is not in time; they have grown incurable. It is the same as with the hectic fever; the physicians say that when the disease begins it is easy to cure but hard to recognize, but in the course of time, when not recognized and treated at the beginning, it becomes easy to recognize and hard to cure. So it is in things of state; on early recognition (which is granted only to a prudent man), the maladies that spring up in a state can be healed speedily; but when, not being recognized, they are allowed to increase in such a way that everybody recognizes them, they can no longer by remedied.[4]

This description of the Romans as doctors and their empire as prone to disease is similar to Plato's discussion of the techno-*polis* from chapter 2. Machiavelli also employs the metaphor of the chronically ill patient to describe the political community and the strong leader as a cure. Here, though, rather than trying to craft a just city and happy citizenry, Machiavelli is more concerned with conquering people and keeping territory. He realizes that because human beings are definitely not political animals, we require a strong dose of "medicine" in order to live peaceably together under common laws and leaders. As he explains, "Because we can say this about men in general: they are ungrateful, changeable, simulators and dissimulators, runaways in danger, eager for gain."[5] Therefore, this remedy must somehow transform naturally selfish individuals into obedient citizens. It is with this in mind that Machiavelli provides a host of means to con, cajole, and coerce individuals to acquiesce to state power. Left to our own devices, free of this cure, the disease will quickly return and we will fall back on our natural instincts, resulting in the fracturing of the state and the death of the patient. Because the Romans were prudent in this way, they were able to gain and maintain a vast empire.[6]

This then is *prudenza* – the capacity to diagnose and treat, to foresee and forestall, the self-interested and malevolent tendencies of human beings. Clearly, this *prudenza* remedy is not concerned with bringing out the best in the citizenry, allowing them fulfil some personal *telos*, but, in a dramatic lowering of expectations, seeks to prevent the worst – political disorder, insecurity, and violence. And while he never actually writes the well-known turn of phrase, this encapsulates the Machiavellian notion that "the ends justify the means." Now it really does not matter if our habits are good or bad, but only if they deliver us a desired end. Machiavelli even states that a prudent prince must "acquire the power to be not good."[7] Now, because the "good" or "God" are no longer targets for our actions, control and manipulation of the world around us and our fellow human beings become the keys to political success and personal fulfilment. In this way, Machiavelli's pragmatic political leader, his Prince, is much like Plato's *technites*-king; only that now, instead of the construction of good citizens, he is interested only in the pursuit of power. Because there is no higher order, first principle, or divine stamp to guide the individual to right judgments, decision making is reconceived as a technical skill that reflects strategy and tactics rather than ethics. This is an important change because it dismisses supranatural or external standards to guide the application of technical knowledge to both nature and human nature. At this point, with no standard beyond the boundless desire to gain and maintain power to guide it, we see technology unleashed from its ethical chains to do whatever we want or whatever it will.

Just as Aristotle's ancient technician imposes form onto matter to arrest the movement of nature,[8] the Prince uses politics to reform and mould individuals to become loyal subjects. For Machiavelli, politics is *techne*. Thus, politics is no longer a practice that is an end in itself, but an external efficient cause for the creation of some other product. He writes, "We see that [prudent leaders] had from Fortune nothing more than opportunity, which gave them matter into which they introduce whatever form they choose."[9] All things considered, this is a remarkable change in attitude. The natural world and all of humanity is reconceived as mere material, waiting for a purpose to be imposed upon it by a powerful and prudent leader.

Of course, this can only be taken so far. While Machiavelli may have set up the massive modern project to reorganize the planet and everything on it to deliver human ends, he thinks that we can only be in command of half of everything. The other half still remains outside our power.

Thomas Hobbes: The Great Reversal

The English political philosopher Thomas Hobbes (1588–1679), however, thinks that we can control all of it. For him, there is no fundamental barrier to our domination of nature and human nature. There is no concept of the good, no counsel from God, and no battle with *fortuna*. There are only technical limitations. The reason why Hobbes thinks we have as of yet been unable to gain this total control is our stubborn reliance on the fallible judgments, the unreliable decision making of human beings. So, while Augustine, Aquinas, and Machiavelli seek to diminish, narrow, and transform the virtue, Hobbes wages an all-out attack on prudence.

Hobbes was the archetypal polymath. Fluent in Latin and ancient Greek, he was a respected translator of Thucydides's *History of the Peloponnesian War*. A student of Francis Bacon, he was an accomplished natural scientist. Hobbes was also a respected inventor, working in the area of lenses and optics. Appropriately, he became friends with Galileo Galilei. But he is best known for his political philosophy. Even in his seventeenth-century English, his ideas sound surprisingly up to date. We more or less agree with Hobbes that scientific facts are superior to the imperfect judgments of friends, neighbours, and colleagues. We agree that the advice of a twenty-first-century *phronimos* (if there even is such a thing) can at best be considered folksy wisdom when compared to the accepted verdicts of the scientific establishment. We likely concur that it is foolhardy and even dangerous to reject clinically proven treatments and therapies for our problems, viewing traditional remedies as antiquated and haphazard.

And yet Hobbes's definition of prudence initially seems quite similar to ancient *phronesis*. He writes, "But this is certain: by how much one man has more experience of things past than another; by so much also he is more prudent, and his expectations the seldomer fail him."[10] Like Aristotle, he understands life experience as the foundation for the development of good judgment. But, unlike Aristotle, he does not hold this capacity in high esteem. According to Hobbes, a prudent man is nothing more than a good guesser: "He that is most versed and studied in the matters he guesses at: for he hath most *Signes* to guesse by."[11] Critical of both ancient and Christian conceptions, for him "signes of prudence are all uncertain; because to observe by experience, and remember all circumstances that may alter the successe, is impossible. But in any businesse, whereof a man has not infallible Science to proceed by; to forsake his own natural judgement ... is a signe of folly."[12] We rely on human judgments *only* when "a man has not infallible Science to proceed by." Now science comes first. Prudence is plainly second rate. What Hobbes is describing then is nothing less than the Great Reversal of the hierarchy between the virtues of *phronesis* and *techne*. The *technites*-king dethrones the *phronimos*. From here on the increasing power of technical knowledge continues to subordinate the weakening authority of human judgments.

This Great Reversal of *phronesis* and *techne* was by no means limited to Hobbesian political philosophy, but spanned the whole of Western civilization and ultimately the entire planet. The Scientific Revolution, including the paradigm-shifting work of Galileo and Newton, provided evidence that the universe was nothing more than a complicated mechanical system that could be rationally understood and explained under a precise set of laws.[13] New techniques allowed for the wider exploitation of nature. Inventions such as the steam engine mechanized labour and drove forward massive industrialization, economic growth, and social upheaval. More and more, the promise of technology rather than the judgments of men pushed the direction of society.

The point is that these revolutions in science and industry would not have occurred without a commensurate change in how we relate to others and the world around us. Because the Great Reversal articulated by Hobbes indicates the new primacy of technical thinking over the judgments of human beings, it follows that life experience simply does not provide anyone with the competence to make proper decisions or understand how the world truly works. Instead, true knowledge requires what Hobbes's old teacher Francis Bacon calls a "skilful minister" that is able to "apply force to matter" and "torture and vex" nature so that it reveals its secrets. Bacon is describing the new scientific method that becomes the foundation for an ever-widening experimentation on the natural world – unlocking the hidden qualities and energies of matter that go on to power the scientific and industrial revolutions.

The novelty of this way of understanding nature is well illustrated by the renowned debate spurred by Galileo's proofs regarding the heliocentric universe. While the controversy stemmed from what was perceived as a conflict between Galileo's findings and sacred scripture, the idea that the earth orbits the sun also defies common-sense experience. Anyone watching a sunrise will be led to believe that it is the sun that is moving. "Ignorance of remote causes," Hobbes decides, "disposeth men to attribute all events, to the causes immediate."[14] Because the "remote cause" of the sunrise is hidden from immediate view, we attribute it to the wrong source. In the same way, everyday living and observation leads us to incorrect conclusions about most of the inner workings of our world and thus provides no grounding in the way things really work. The *phronimos* merely can guess, whereas the scientist actually knows because only he is able to unearth the hidden laws of nature.

The same is true of politics. Rather than understanding the remote causes of human behaviour, politicians in Hobbes's day merely could guess what goes on the minds of their citizens. He advised that experts who understand humanity on a technical level should instead run government. He calls upon "good counselors" to use their expert knowledge to design bureaucratic institutions that will allow the state to run like an automatic mechanism or machine. At first, these modern political scientists, these counsellors, bear a similarity to a *phronimos* because they require "a great knowledge of the disposition of Man-kind ... And the Strength, Commodities, Places both of their own Country, and their Neighbours; as also of the inclination, and designes of all Nations."[15] But, unlike an education in ancient political science, all of this experience and training is not directed towards the development of a good leadership, practical wisdom, or even good citizens. Instead, Hobbes expounds the importance of this experience and training because it leads to a grasp of an administration based on "Infallible rules, (as in Engines, and Edifices, the rules of Geometry,)."[16] He continues: "All the experience of the world cannot equall his Counsell, that has learnt, or found out the Rule." So, it is not the particular life of the good counsellor that makes him an invaluable element of the state but his or her scientific knowledge or understanding of the "infallible rules."

Just as the natural scientist is able to reveal and exploit the hidden qualities of the natural world, Hobbes argues that this new kind of political scientist can do the same with the political world. Again, the obstacle has been our stubborn reliance on the very flawed virtue of prudence. If an exact technique of politics could be discovered to replace it, we would no longer have to rely on the whimsical decisions of the political elite nor would we have to contend with the unpredictable movements of the masses. So, prudence actually presents Hobbes with two related problems. The first is the ignorance of political leaders and the irrational judgments they make based on their less-than-reliable life experiences. The second and even more perplexing problem is the varied judgments of the citizenry. Because prudence is derived from a diversity of personal experiences rather than universal rules of reason, no two people can be expected to have the same reaction or make the same decision in any given circumstance.[17] Prudence not only makes for bad leaders but also makes it nearly impossible to predict and control the thoughts and actions of people. There are simply too many contingencies to deal with. Recall that Aristotle thinks a purely technical knowledge of politics is implausible due to the sheer complexity of human beings. Even Machiavelli demurs at the bold claim that a prince could completely predict and control human nature. Hobbes decides that this is not insurmountable truth but merely an assailable technical limitation.

He recognizes that a science of human management will not be easy. But he remains confident that just as the exact rules of geometry were uncovered through hard work and study, so too will the exact rules of politics, although he admits, "Politiques is the harder study of the two."[18] If the rules of politics were structured like the rules of geometry or physics, political life would be predictable and not subject to turmoil and violence. In his new age of discovery, Hobbes asserts that just as other long-held problems and superstitions had been solved or dispelled, so too could this enduring problem. It is just a matter of time and industry:

> Time, and Industry, produce every day new knowledge. And as the art of well building, is derived from Principles of Reason, observed by industrious men, that long studied the nature of materials, and the divers effects of figure, and proportion, long after mankind began (though poorly) to build: So, long after men have begun to constitute Common-wealths, imperfect, and apt to relapse into disorder, there may, Principles of Reason be found out, by industrious meditation, to make their constitution (excepting by externall violence) everlasting.[19]

In this extraordinary passage Hobbes not only compares statecraft to the technique of building, but he also suggests that, if constructed according to the "Principles of Reason," his commonwealth would be

perfect, held together with a "constitution everlasting" that will perpet-
ually resist the internal forces of disorder that have ripped apart every
political community hitherto. Just as the house builder imposes the
form of a house onto his materials, the state builder must impose the
form of the state onto human beings. But, whereas even the best built
house will not last forever, Hobbes seeks to create something that will.

For the modern state builder, the key is to fully understand the nature of his materials: human beings. Aristotle's earlier reservations
about the portrayal of humans as mere *choregia* or "human material"
stemmed from the false belief that human beings living together in a
polis would be too difficult to control in a technical way. In turn, the
polis in the *Politics* was "imperfect and apt to relapse into disorder," a
product of an earlier ignorant age. But, because man has now accumulated enough new knowledge through "industrious meditation," this
ignorance can be pushed aside and the perfect political community can
finally be made. This is why Hobbes colours all earlier political traditions as simply inferior, less-informed versions of his present project.

Aristotle's mistake was to think that the role of the state was the fulfilment of the citizen's personal *telos*, bringing him to the final end of *eude-
monia*. Instead, Hobbes concludes that all humans are fundamentally
equal because they share a "similitude of the passions" or sameness
of the senses. Because this equality allows for a common perception of
pain and pleasure, the sovereign state or "visible Power" can effectively
control every citizen through the threat of punishment:

> The finall Cause, End, or Designe of men, (who naturally love Liberty, and
> Dominion over others,) in the introduction of that restraint upon them-
> selves, (in which wee see them live in Common-wealths,) is the foresight
> of their own preservation, and of a more contented life thereby; that is
> to say, of getting themselves out from that miserable condition of Warre,
> which is necessarily consequent (as hath been shewn) to the naturall Pas-
> sions of men, when there is no visible Power to keep them in awe, and
> tye them by feare of punishment to the performance of their Covenants.[20]

Rather than pleasure or happiness, the tangible fear of pain, our mutual
"Fear of Death, and Wounds,"[21] is the glue that holds the modern nation
state together. We are not obliged to follow the laws of the state or per-
form our "Covenants" out of a sense of virtue or loyalty but at bottom
out of our common sense of self-preservation. Like trained rats in a scientist's laboratory, we can be shocked and prodded to move one way or

the other, following this law and avoiding that prohibition. Hobbes takes this idea a step further to argue that "the Common-peoples minds ... are like clean paper, fit to receive whatsoever by Publique Authority shall be imprinted in them."[22] So, the modern state builder not only is able to control how we act, but also seeks control of what we think. In turn, Hobbes promotes propaganda and censorship as well as the silencing of "vainglorious" challengers to power as familiar and effective methods for the consolidation of the state's authority over the citizenry.

In order to succeed, all the unpredictable passions, desires, and irrationalities of human beings must be quantified and controlled. In chapter 7 of *Leviathan*, Hobbes worries about an "excess of passion" that pushes men to seditious thought and actions. And this worry leads him to put all citizens under suspicion of revolt. As he describes, "For the Thoughts, are to the Desires, as Scouts, and Spies, to range abroad, and find the way to the things Desired."[23] From this we can gather that Hobbes takes the passions to be a clandestine source for our behaviour. It is the task of a good government to locate and contain any dangerous elements (i.e., "Scouts and Spies") of human thought and action that could potentially disrupt the proper running of the state. In all, the state is directed towards the manipulation of human nature through greater scientific understanding of the passions and the implementation of that understanding in the design of laws and public institutions. For Hobbes, they are the instruments of external control. Rather than completing human nature, politics is concerned with manipulating the appetites. As he explains: "But as men, for the atteyning of peace, and conservation of themselves thereby, have made an Artificiall Man, which we call a Common-wealth; so also have they made Artificiall Chains, called Civill Lawes."[24] The better these laws and institutions control and limit our natural inclination, the greater opportunity there is for domestic peace and stability. So, from the very start of *Leviathan*, Hobbes sets us off on a course wholly different from the understanding presented by the classical thinkers. In disregarding the possibility that the passions lead us towards a natural and teleological good, Hobbes dismisses Aristotle's description of the hierarchical relationship between the appetites and the virtues. For Hobbes, the state manipulates our nature through artifice. Where in the classical conception virtue moderates the appetites, in Hobbes all we are is appetites.

Because the specific objects of pleasure differ so much between men, Hobbes bases the manipulation of the appetites on our common aversion to pain (i.e., "Fear of Death, and Wounds").[25] As it is based on an alleviation of pain and not a fulfilment of pleasure, politics is cut off from satisfying the passions, or is de-eroticized. The desire for pleasure, the initial spur for all activity in classical thought, is confined, held down, limited to acquisition and wealth. The *Leviathan* then is something of an instruction manual for the modern practice of kingly *techne* or a blueprint for the ultimate building of the techno-*polis*. Of course, just as Plato's city seemed abhorrent, Hobbes's insertion of political control into the intimate recesses of our minds and bodies makes his political philosophy seem not the inspiration for the modern nation-state in general but rather the Orwellian realms of Stalin's Soviet Union or Mao's China.

But unpredictably, Hobbes actually baulks at the idea of a totalitarian technical control of humanity. In what seems a major step down, he concedes that, "For seeing there in no Common-wealth in the world, where in there be Rules enough set down, for the regulation of all the actions, and words of men, (as being a thing impossible:) It followeth necessarily … by laws praetermitted, men have the Liberty, of what their own reason shall suggest, for the most profitable to themselves."[26] Rather than total control, he seems to accept that as long as individual citizens follow the stated laws, they will and should remain at liberty to pursue whatever pleasures and satisfy whatever appetites they want in their *private* lives. So, while politics is purged of the unreliable judgments of human beings, we remain free to make our own (often irrational) choices outside of the public realm. This again makes Hobbes sound quite up to date in that most citizens today spend their time pursing personal economic goods with little concern for the common good.

Still, it is not clear why at this point Hobbes claims that it is "impossible" to have a commonwealth that completely masters the actions and words of its citizens in both their public *and* private lives. Is this simply the consequence of some remaining technical limitation? This seems the most likely answer. Because it so difficult, a complete and proper method of "Politiques" may require something still beyond the knowledge of men: "Neither Plato, nor any other Philosopher hitherto, hath put into order, and sufficiently, or probably proved all the Theorems of Morall doctrine, that men may learn thereby, both how to govern, and how to obey; I recover some hope, that one time or other, this writing of mine, may fall into the hands of a Soveraign, who will … convert this Truth of Speculation, into the Utility of Practice."[27]

He holds out the hope that a statesman may someday come to this full understanding and apply a complete science of society.[28] But, is Hobbes actually calling for a completely mechanical, and thus predictable, understanding of humans or is he instead arguing that humans maintain something uncertain, outside of mechanical explanation?

Only because politics has yet to be understood in the same way as physics or geometry does it require a different approach that includes consideration of ambiguous human subjectivity, desires, and rights. That is to say, rather than being able to fundamentally alter human nature, the state must keep a vigilant watch for insurrection against state power. Arguably, this is only for a lack of technical expertise. If a blueprint of human thought and psychology could be constructed, then a purely objective approach to the citizenry would suffice, as they would be nothing more than malleable objects or material. And as it is left at the end of his description of his new, modern commonwealth, Hobbes waits for that day. It is only in the next century, during the period known as the Enlightenment, that we see that day arrive.

5 After the Great Reversal

This chapter is explores the Enlightenment as an outcome of the rise of unlimited technical knowledge freed from prior limiting capacities. By applying a "science of society" model to every aspect of human life, Enlightenment thinkers not only transform the natural world into a resource for progress and development, but also similarly transform human nature. The body and mind are reconceived as malleable matter in need of remoulding towards a new and better version of man. Good judgment, based in particular experience, is fully subordinated to this universal modern project to perfect humanity. In the wake of the Great Reversal, we finally see the full impact of forgetting Aristotle's warning that in a technically run *polis* society human beings become ever more worked upon as mere material and how this kind of treatment bars the practice of *phronesis*. So, this period not only sees the foundations of our contemporary technological society being laid, but also sees our ability to understand and criticize it constrained.

Hobbes's Challenge

Hobbes's convincing rejection of the classical and Christian traditions cleared the way for the practice of a new kind of politics. Instead of the good or God, politics could now be based on "infallible rules" that govern human behaviour. If these underlying rules could be exposed and exploited to manipulate the passion of the citizenry, governments could function under a "constitution everlasting" and no longer be threatened with unpredictable crises and civil unrest. Hobbes's challenge to the next generation of political philosophers was to locate a specific inventory or instruction manual of these long-sought-after rules of man.

The eighteenth century, part of the period known as the Enlightenment, was to a great extent occupied with this challenge. Of course, where Hobbes's main concern was the application of the "infallible rules" to government, the Enlightenment thinkers went further, seeing an opportunity to apply them not just to public life but to all areas of human conduct: not just politics but also economics, social institutions, ethics, and human relationships in general. They sought out a definitive science of man, a moral science, or a "social science" that would overcome the ignorance, social chaos, intolerance, and inequality that were the cause of historic oppression and suffering. With tremendous hope and brimming with new confidence, they thought that this science of society would eliminate the long-held irrationalities and petty differences that divided humanity, finally putting aside the archaic tribal and religious loyalties that had fuelled pointless conflict and war for century upon century. For the scientists and philosophers of this era, true happiness and everlasting peace were within reach. A time of extraordinary creativity and upheaval, the Enlightenment is a first glimpse into life and society after the Great Reversal – the long-awaited result of the technical vision of life and society unleashed from the ancient bonds of flawed human judgments and superstitions.

This optimism was in good part spurred by the amazing strides made during the Scientific Revolution in our understanding of the make-up and function of the natural world. For one thing, our newfound comprehension of nature greatly increased our ability to utilize natural resources and dramatically improved agriculture yields, all of which provided for a huge expansion of the economies of Europe. In the New World, the embrace of Enlightenment values pushed the American frontier west to tame the wilderness and provided the material basis of a burgeoning technological society.[1] Now, all of nature is viewed as a massive untapped resource. The rivers, lakes, forests, fields, mountains, and animals are ready to be harvested as generators of new wealth and power.

As significant were remarkable discoveries about the function of the human body that would have a direct impact on the health and well-being of ordinary people. The new natural science that had given so much insight into non-human nature a century earlier could now be applied with equal success to human nature. The muscles, the skeleton, the organs, the nerves, the blood, every body part and drop of fluid, were subject to intense inquiry and research. In laboratories and scientific academies all across the continent, new breakthroughs were being made that led to cures and treatments for common illnesses that used to kill people by the millions. These breakthroughs run from the celebrated, such as Antonie van Leeuwenhoek's observations of microorganisms, which led to the founding of microbiology, to the obscure, such as the Scottish scientist Edward Stevens's work on the stomach's digestive juices.[2] Overall, this work eventually led to huge improvements in the way people lived. There were major advances in hygiene, new ideas about the origins of disease, the monumental discovery of inoculations and vaccines, the manufacture of new medicines, a growing comprehension of human health, and the rise of modern medical science itself.[3]

In tandem with this success in quantifying and controlling the human body, there was also an attempt to quantify and control human behaviour. Various grand theories on the development of a social science as a counterpart to the natural and medical sciences were put forward by many of the Enlightenment's most significant thinkers, including Blaise Pascal, Gottfried Leibniz, and later David Hume as well as the French philosophers Montesquieu, Condillac, and Auguste Comte. All of them agreed that, because human behaviour was ruled by the same set of universal laws that governed the natural world and human physiology, identifying these laws would grant them the power to effectively predict and control social movement and organization in the same way natural scientists and physicians had already learned how to manipulate raw materials and manage human health.

However, a fundamental problem remained. The development of a corresponding "social science" is predicated on the idea that the "infallible laws" of human behaviour were in fact truly *universal* and applied to all humans equally. As Hobbes had already pointed out, while humans may be born equal, the diversity of their particular experiences created unmanageable differences between them. These difference are what Jean Jacques Rousseau (1712–78) thought led to social inequality, which was later legitimized and enhanced by unjust political and social institutions. The burgeoning field of social science saw the solution to this stubborn problem as twofold. First, in order to mitigate the diverse circumstances of individuals, decision making had to be wrestled away from experience. Second, the actual content of experience had to be managed so that it provided a common foundation upon which a truly equal society could be built. Together, this would not only provide the groundwork for universal, commonly shared, and agreed upon ethical and political judgments, but would also drive a reordering of society to rid it of entrenched institutions that perpetuated inequality.

The leading philosopher of the day, the German idealist Immanuel Kant (1724–1804), went a long way to cutting judgment away from its roots in experience. If one understands the Enlightenment as an effort is to purge any and all irrational elements from everyday life, to live a life of reason rather than superstition or religion, tradition, or custom, it follows that individual decision making is also to be reconceived as a product of rational thinking rather than a reflection of the rough ground of everyday life. Kant describes judgment, not as an expression of the hurly burly influence of parents, role models, and particular life experiences, but as a calculus for the maximization of happiness or, as one commentator puts it, a "computer which needs a program for deciding moral questions."[4] Much like Aquinas, Kant thinks good judgment is a "top-down" affair, informed by universal rules or fundamental laws discovered through reason. In his *Groundwork of the Metaphysics of Morals*, he flatly rejects the idea that experience can in any way inform morality, explaining that "we cannot do morality a worse service than by seeking to derive it from examples. Every example of it presented to me must first be judged by moral principles in order to decide if it is fit to serve as an original example – that is, as a model: it can in no way supply the prime source for the concept of morality."[5]

Remarkably, Kant does not simply dismiss the place of everyday life in the development of our ethical sensibilities, but decides that "we cannot do a worse service" than to found ethics on our own particular experiences. Not only does this mean we cannot trust our own feelings of satisfaction or pleasure as informing good decision making, but also suggests we can learn absolutely nothing from our particular upbringing, friends, and colleagues. For Kant, exemplars and role models are more than a waste of time, but actually muddy our thinking, acting as bad influences that mislead us to make wrong decisions.

Kant instead tries to universalize prudence to make it available to every individual regardless of their circumstance and foolproof in every situation, disconnecting it from random influences. He wants to do away with the guesswork of making right choices, replacing it with an infallible equation that will always provide the proper solutions to even the toughest ethical dilemmas. His celebrated *categorical imperative* advises individuals to "act only according to that maxim by which you can at the same time will that it should become a universal law."[6]

He wanted to blind or numb the decision maker to the particularities of his circumstance and asked them to instead universalize their choices as though they were making a decision for all of humanity. For Kant, the well of good judgment should no longer be tainted by the toxin of everyday life.

But, this is only half of the solution. The radical rejection of experience, this turning away from experiential knowledge, also demands a drastic reordering of society to conform to universal laws that govern human behaviour. This effort is given a boost by extraordinary advances in mathematics that led to the development of original areas of inquiry, including the new field of statistical analysis. Suddenly, government officials had the ability to account for the complexities of population movements and densities and keep track of mortality rates and birth rates. This offered them a powerful new tool to calculate future economic and social conditions and allowed for the implementation of new policies in an anticipatory manner rather than merely reacting to problems as they occurred. Even though we now use statistics to measure an incredibly wide range of things, it was first conceived specifically as a science dealing with data about the condition of a state. It could be said that statisticians represented a totally new breed of public servant that, in a way never before thought conceivable, had a scientific knowledge of the function and make-up of whole populations. What was before viewed as a swirling, chaotic mass of humanity suddenly became a complex logical system that could be reduced to points of data laid bare on a mathematical table. It should come as no surprise that the power of this information helped spur a massive expansion of state bureaucracies all across the countries of Europe. Through the work of these bureaucrats, all the vagaries, unpredictability, and chance formerly associated with political life could be ordered and tamed once and for all.

Basically, this new ability to quantify mass social phenomena finally supplied governments with a long-sought-after technique to control their citizenry. Centuries earlier, in his great work defending the foundations of the French kingdom during a time of prolonged civil war, the *Six Books of the Commonwealth* (1576), the political philosopher Jean Bodin anticipated an increasingly important role for a more precise use of census data beyond the tradition of counting citizens available for military service or property owners for tax collection. Bodin explains that an idea of the number and distribution of the population ... makes it possible to get rid of those parasites which prey upon the commonwealth, to banish idlers and vagabonds, the robbers and ruffians of all sorts that live among good citizens like wolves among the sheep. One can find them out, and track them down wherever they are ... Again, drunkenness, gambling, fornication, and lust can be indulged in without check from the law. Who can remedy this state of things but the censor? One sees also how most commonwealths are afflicted with vagabonds, idlers, and ruffians who corrupt good citizens by their deeds and their example. There is no means of getting rid of such vermin save by the censor.[7]

According to Bodin, not only would this data assist in the making of good social policy, but it would also help root out undesirable elements that "corrupt good citizens" and, in turn, help ensure the proper education, virtue, and morality of the population as a whole. Like a doctor diagnosing the cause of an illness, the censor or statistician is charged with identifying "such vermin" so that they can be purged from the system.[8]

Bodin's hope for this kind of aggressive social engineering becomes a reality with the establishment of new agencies charged with compiling and analysing population data. In England, William Petty (1623–87), a friend and colleague of Hobbes, developed what he called "political arithmetic," allowing the government to make more accurate calculations of national wealth. In mid-eighteenth century France, the Bureau of General Statistics began for the first time to collect accurate data sets covering every parish and farmhouse across the French countryside. It was hoped that this information would also allow for the government to identify areas of social unrest and adapt their policies accordingly.

The French mathematician and philosopher Condorcet (1743–94), mentioned in chapter 1, was perhaps the leading supporter of the predictive power of the statistical analysis of human behaviour. However, he saw it as more than simply a tool to measure and manipulate social and economic organization, but also as a means to fundamentally reorder society along more rational lines. In his final, great work, written while in hiding and published after his suicide in a Parisian prison cell, *Sketch for a Historical Picture of the Progress of the Human Mind*, he writes in a characteristically buoyant fashion that "nature has set no limit to the realization of our hopes" and that "the perfectibility of man is truly indefinite." His thinking on the progress and perfectibility of man led him to wonder:

Would it be absurd now to suppose that the improvement of the human race should be regarded as capable of unlimited progress? That a time will come when death would result only from extraordinary accidents or the more and more gradual wearing out of vitality, and that, finally, the duration of the average interval between birth and wearing out has itself no specific limit whatsoever? No doubt man will not become immortal, but cannot the span constantly increase between the moment he begins to live and the time when naturally, without illness or accident, he finds life a burden?[9]

Like many of his contemporaries, the possibility of quantifying and controlling the actions of men filled Condorcet with a boundless optimism. His earnest belief in the unlimited progress of human beings, including the extension of human life beyond any specific limit, epitomized the prevailing attitudes of the era. All limits on human progress, whether disease and violence or poverty and ignorance, were thought to be eliminable through the application of reason to the world.

It is from this premise that Condorcet confidently develops the field of "social mathematics," designed to provide a predictive model of human behaviour. No longer would populations have to be managed through a combination of coercion and guesswork, but instead through the precision offered by equations and formulas. In an echo of Hobbes's emphasis on the "infallible rules," he thinks this engineering of populations is feasible because "the truths of the moral and political sciences are susceptible to the same certainty as those forming the system of the physical science, even those branches like astronomy which seem to approach mathematical certainty."[10] He takes this idea even further, arguing that the processes of the human mind themselves are controllable on the same ground, and pointing out that "the sole foundation for belief in the natural sciences is this idea, that the general laws directing phenomena of the universe, known or unknown, are necessary and constant. Why should this principle be any less true for the development of the intellectual and moral faculties of man than for the other operations of nature?" So, it was thought possible to quantify not only the external organization of populations but also the inner workings of individuals. According to Condorcet, this deep application of social mathematics would require not only a drastic reordering of human society but also a fundamental shift in the way children were reared and educated.[11]

The idea, of course, was to finally overcome the superstitions and traditions that generated the social inequality and backwards thinking that remained stubborn barriers to further progress. But, as Condorcet had tragically learned, there was a starkly different approach to solving these problems as well as a different idea about who were the real "parasites" on the body politic that was gaining strength in the streets and back alleyways of Paris. Just as scientists and philosophers embraced the power of reason as the saviour of humanity, the French revolutionaries sought out the complete eradication of the right of kings and queens to claim authority over men based on an irrational and indemonstrable relationship to God. The hierarchies that sustained political power and enforced law for generations, what Hobbes had called the "invisible power" of divine right, were to be replaced with new, rational systems of government. As the spirit of the French Revolution swept across the continent, it became clear that the road to true happiness and everlasting peace would be far from happy or peaceful.

Perhaps the early optimism of Enlightenment thinkers blinded them to the incredible release of violence that was about to wrack Europe and much of the world. The initial hope of Condorcet and others was that there could be a calm transition from the old standard to the new. And, about fifteen years before the revolution, Condorcet's mentor, the leading French bureaucrat and economist Anne Robert Jacques Turgot (1727–81), was prescient in his recognition that French society was under tremendous strain from an outdated system of government and an inefficient economy. He worried that the disorganization and unpredictability that marked the rule of monarchy was beginning to pull the country apart. In response, he proposed that the newly instated king, Louis XVI, move away from the irrational tradition of "rule by decree" to a system of government and economics based on "general laws" equally applicable to all the citizens of France. Turgot was no ordinary pencil pusher, but was among an elite set of French thinkers known as the *philosophes*. This group of great literary, philosophical, and scientific minds included Voltaire, Montesquieu, Diderot, d'Alembert, and of course Turgot's good friend and brilliant protégé Condorcet. With common conviction, they argued for the creation of a rational, secular, and open-minded society. And so Turgot wrote his letter to the king hoping to convince him that a new rational system of government was needed to overcome the growing internal factions developing within France and to create a prosperous and unified nation:

The cause of the evil, sire, stems from the fact that your nation has no constitution. It is a society composed of different orders badly united, and of a people among whose members there are but very few social ties. In consequence, each individual is occupied only with his own particular, exclusive interest; and almost no one bothers to fulfill his duties or to know his relationship to others. As a result, there is a perpetual war of claims and counterclaims, which reason and mutual understanding have never regulated, in which Your Majesty is obliged to decide everything personally or through your agents. Everyone insists on your special orders to contribute to the public good, to respect the rights of others, sometimes even to make use of his own rights. You are forced to decree on everything, in most cases by particular acts of will, whereas you could govern like God by general laws if the various parts composing your realm had a regular organization and clearly established relationship.[12]

Here again we see a call for the development of a social science that could finally replace the outdated and unwieldy practice of a rule by decree based on the monarch's flawed and erratic judgment. With everyone treated equally under rationally designed social and political institutions, Turgot reasoned, French society could be ordered into a predictable system that would be able to address the growing dissatisfaction and unrest across the kingdom. Unfortunately, his letter never reached the king. Having fallen out of favour with the royal family and been forced to resign, Turgot died before his concerns were borne out in the violence of 1789.

But, it is not as though the Revolution put an end to such ideas. Far from it, the triumphant revolutionary government accelerated the implementation of rational standards in all aspects of French society. A universal metric system was adopted to standardize weights and measures. The calendar was changed to restart at Year I, representing the first year of the new French Republic. The months, weeks, and days were renamed and reordered to reflect an emphasis on mathematical precision. Time was now to be gauged in decimals of ten, allowing for ten hours in a day, each hour composed of one hundred minutes and each minute composed of one hundred seconds. Perhaps the strangest change was the introduction of the *Culte de la Raison*, or Cult of Reason, as a new religion to replace backwards Catholicism, with the Goddess of Reason as the focus of parishioners' worship.[13]

Eventually, the Republican calendar, decimal time, and the *Culte de la Raison* fell into disuse or disrepute. But the Enlightenment ideals that inspired them would spread from country to country and have a significant influence on the development of the new government across the Atlantic in America.[14] The combination of natural science, medical science, and social science effectively helped bring about a new approach to understanding and controlling nature, the human body, and society.

For some philosophers, the series of amazing discoveries and events that marked the eighteenth century suggested that this era was more than just a sum of its parts. They argued that what we were seeing was in fact the coming to fruition of history itself and that humanity had finally progressed to a perfect end of history. With enough information or data, they thought, one could scientifically calculate and locate universal rules for each and every human action and thought. Because all individuals are the same kind of rational animal one can catalogue not only human physiology,[15] emotions, and psyche, but also human politics, social relations, economics, and ethics. Under the power of these scientific accounts, institutions could be set up with little concern for the random life experiences of citizens and leaders or the particularities of time and place.

Thus, we can understand the Enlightenment as not only Hobbes's challenge fulfilled but also a further extension of his project both inwardly and outwardly. The Enlightenment philosophers' efforts to create a science of society designed to perfect man were a deeper application of the early modern attempt to cure the selfish and violent character of human nature through the building of a scientifically managed state. The promotion of a universal rationality that nullifies any and all disagreements between men was part of the same plan as Hobbes's project to control the complexities of individual citizens and communities. In both cases, there is an increasing technical control of human life and a replacement of traditional institutions by a universal society governed by scientific principles. Where Hobbes wished to purge ignorance and irrationality from public life, these eighteenth-century thinkers push the dictates of reason and science into the innermost recesses of the private sphere.

This is a significant turn in modernity. The Great Reversal brought in a new era where technical knowledge came to dominate human judgment. But, during the Enlightenment, there is a further unprecedented acceleration of science and technology into all areas of human life. Now, instead of good judgment guiding technical knowledge, technical knowledge comes to guide judgment, turning the ancient virtue of *phronesis* into a subordinate of a larger scientific project to perfect humanity.

Following the optimism of the eighteenth century, the nineteenth century is taken up with an effort to see the social sciences finally catch up with the flurry of successes achieved in the fields of natural and medical sciences. The Belgian mathematician Lambert Adolphe Quetelet (1796–1874) attempted to pick up this effort where Turgot and Condorcet left it off. As with his predecessors, Quetelet's goal was to identify the underlying laws governing social phenomena and develop new social policy and institutions to give governments a more complete control over their population's health and well-being. The idea again was to eliminate the sway of the accidental and arbitrary on the movement of human progress. In lectures published in 1828, Quetelet explains that what might seem like random events affecting everyday life are actually part of a larger complex system:

> Chance, a mysterious word, which has been too much abused, ought only to be considered as serving to conceal our ignorance; being a phantom which exercises a most absolute empire over the vulgar mind, habituated to consider events only as insulated facts; but which are annihilated before the philosopher, whose eye embraces a long series of events, and whose observation is not deceived by irregularities, which disappear from before his steady view, when he is enabled to take a position sufficiently elevated to sieve the law of nature.[16]

Chance, Machiavelli's *fortuna*, is no longer a worthy opponent but is reduced to a "veil for our ignorance" that can be lifted through the precise study of the events, variations, and laws governing the average man. To this end, Quetelet goes a step further than Condorcet's "social mathematics" to design a "social physics" that attempted to highlight the interrelation of physical, social, intellectual, and moral development of the "average man." His 1835 *A Treatise on Man and the Development of Human Faculties: An Essay on Social Physics* makes clear that what seemed "insurmountable" complications associated with an inquiry into the causes behind the human will could now be discovered by taking a wide perspective,

> It is in this way that we propose studying the laws which relate to the human species; for, by examining them too closely, it becomes impossible to apprehend them correctly, and the observer sees only individual

peculiarities, which are infinite ... It would appear, then, moral phenomena, when observed on a great scale, are found to resemble physical phenomena.[17]

Quetelet realized that there are many "active causes" that have a deep influence upon the function of society and that they are not quickly or easily purged from the social body. However, he also recognized that man can in some measure master these causes and "improve his condition." He writes that, "in order to succeed, we must study the masses, with the view of separating from our observations all that is fortuitous or individual."[18] In the same vein as Kant's rejection of experience, Quetelet instructed social scientists to also discount anything individual or isolated. And while he agreed that, just as a mortality table will not help predict the date of death of any one individual, the general laws discovered through such scientific research would nonetheless help legislators govern the social body in general. He even admitted that average man is a "fictitious being" that merely represents the "medium results for society in general."

This idea was a great inspiration to the English statistician Francis Galton (1822–1911). Quetelet's work on the fictitious average man convinced Galton that the physical, intellectual, and moral characteristics of human beings were affected by the "disturbing action" or "secular perturbations" of society as much as they were the result of natural selection, an idea his cousin Charles Darwin had recently introduced to the world. He worried that the tremendous growth of government-run social services that aided the infirm, educated the feeble-minded, and benefited the impoverished was disrupting the normal evolutionary development of man by unintentionally allowing undesirable characteristics to be passed down by parents who had previously been unable to live long enough or well enough to have offspring.

In the preface to his 1869 *Hereditary Genius*, Galton explains that "the theory of hereditary genius ... has been advocated by a few writers in past as in modern times. But I may claim to be the first to treat the subject in a statistical manner, to arrive at numerical results, and to introduce the 'law of deviation from an average' into discussions of heredity." In what followed, Galton showed that, just as particular physical characteristics such as height and longevity become more pronounced in certain groups over generations, so too did the attribute of intelligence. Heretofore, these changes were not viewed as falling under the control of human beings. However, by identifying and promoting certain characteristics in human populations, Galton now believed, governments could help guide the future evolution of humanity:

> The entire human race, or any one of its varieties, may indefinitely increase its numbers by a system of early marriages, or it may wholly annihilate itself by the observance of celibacy; it may also introduce new human forms by means of the intermarriage of varieties and of a change in the conditions of life. It follows that the human race has a large control over its future forms of activity.[19]

Through the lens of statistical analysis, Galton pressed governments to test young men and women for desirable hereditary characteristics and then encourage them to marry and have children, thus advancing the human race. In an earlier work, he even lamented: "If a twentieth part of the cost and pains were spent in measures for the improvement of the human race that is spent on the improvement of the breed of horses and cattle, what a galaxy of genius might we not create! We might introduce prophets and high priests of civilisation into the world."[20]

Galton, of course, was a leading advocate of eugenics (from the Greek *eugenes* meaning good in stock or hereditarily endowed with noble qualities). Rather than spending time and money to overcome social injustices through reforming working conditions, health care, educational institutions, human rights, and laws, the eugenicists argued that social progress could be more effectively achieved through selective breeding. Moreover, with human nature and evolution firmly under the control of human beings, Galton also believed, we could then go further to create a "higher order of personality" or superior species of man. The "average man" could now be quantified and controlled.

Clearly, this is a disturbing outgrowth of Enlightenment thinking. But Galton is no monster. He was motivated by the very same humanitarian impulses as his intellectual predecessors, Turgot, Condorcet, and Quetelet, to move humanity forward and to perfect man. Eugenics is just another expected step in a larger effort to completely purge chance from human life and society. In fact, before eugenics grabbed his interest, Galton was a pioneer in the field of meteorology. He realized that changes in human populations over successive generations could be charted in the same way meteorologists were able to quantify or "mathematize" complicated weather patterns.[21] The only difference between meteorology and eugenics was that the statistics gathered on populations could later be used to change the object of investigation. Eugenics then typifies the technological impetuses to quantify and control in a most fundamental way.

The point here is that the period after the Great Reversal is dominated by technical knowledge or the technological vision. *Techne* becomes the lens through which we see and understand the natural world, the human body, and politics. Plato, as we may recall, anticipated in *The Statesman* that the rise of kingly *techne* would bring with it the formation of a "science of government" that seeks to control the composition and temperament of the citizenry. So, despite the millennia that separate Plato and Galton, nineteenth-century eugenics is a rather unsurprising consequence of this way of thinking. The struggle between the "two cities," between kingly *techne* and *phronetic* rule, seemed to be over.

The twentieth century saw an exponential growth of the technological impetuses. The planet and everything on it is subject to quantification and control – every plant and animal, every patch of soil and ounce of mineral, every function of the body and every process of the mind, and even the weather fall under a now dominant technological management.

6 Responses to the Great Reversal

Having now largely completed the first objective of the book to outline the movement away from Aristotle's warning towards the Great Reversal of the judgmental and technical, this chapter returns to the work begun in chapter 1 in an effort to fulfil the second part of the second objective of the book, which is to consider how, in a technological society where human beings are treated as mere material, the practice of *phronesis* may be obstructed or barred.

As will be explained, the return of *phronesis* into the discourse is in large part spurred by a growing unease with the goals and consequences of the Enlightenment. The violence of the twentieth century made it all too clear that the Enlightenment project for a science of man and society did not bring us "everlasting peace" – at least not yet. While the project made great strides towards quantifying and controlling both human and non-human nature, the perfectibility and happiness Turgot, Condorcet, Quetelet, and Galton promised seemed only a distant dream of a past and naive age. If nothing else, the events that marked the first half of the twentieth century told us that the power of technology could as easily be turned to unprecedented destruction as it could to human flourishing. Free of an ethical compass to guide it, technology revealed itself as a potent amoral force.

But, even before the two world wars, the Holocaust, and the Great Depression, there was already a strong undercurrent of "counter-Enlightenment" thinking that pushed back against the early goals of the burgeoning technological society. The "dark Satanic Mills" that William Blake saw as a blight spreading across the early-nineteenth-century English countryside epitomized a growing anxiety that something intrinsic was being lost with each turn of a water wheel and each stroke of a steam-driven piston. In protest, poets and philosophers alike began to express a general worry that modern society as manifest through technology somehow was impeding or corrupting their capacity to live full and complete lives, and barring humanity from experiencing the world in same way as their pre-technological brethren.

Anti-Technology

As mentioned earlier, the idea that humans had become imprisoned by modern society and technology resonated in the anti-Enlightenment Romanticism of Thomas Carlyle, John Ruskin, the poets Lord Byron and Percy Shelley, and of course Mary Shelley, author of *Frankenstein*. On the other side of the Atlantic, the American transcendentalists Henry David Thoreau and Ralph Waldo Emerson expressed the same concerns.[1] A common theme in all of these writings is that, while we may think we are directing technology towards the fulfilment of human-centred goals, technology is all the while forcing us to conform to its demands, and actually using humanity as a tool for its further advancement. This anti-technology sentiment was not limited to poetry and philosophy, but also worked its way down to the people who actually laboured in those same dark mills. The legendary nineteenth-century Luddites, textile workers who feared unemployment through the automation of their industry, directed their antipathy for technology by attacking and destroying the wool-spinning machines that would, despite their efforts, come to replace them.

This alternative tradition of counter-Enlightenment thinking continued into the twentieth century, becoming a sustained and more direct criticism of technology and its negative effects. In Germany especially, the devastation of the First World War, coupled with the global economic depression, saw the rise of a great wave of discontentment with modern life. More and more, the optimism of the Enlightenment gave way to profound disapproval matched by calls for a renewal of older ways of life that had been disrupted by the "quest for certainty" and "conquest of nature" that dominated the centuries after the Great Reversal. Conservative writers such as Ernst Jünger and Oswald Spengler, the author of the widely popular *The Decline of the West* (1918–23), tapped into this mounting sense of dissatisfaction and frustration with their sweepingly pessimistic ideas about the imminent decline of Western civilization at the hands of the out-of-control advance of technology. In

1932, Jünger lamented the spiritual void created by a "cult of progress" that "smashes even traditional ways of life." He blamed the overwhelming and "enchanting" influence of technology, writing that "wherever man falls under the spell of technology, he finds himself placed before an unavoidable either/or. This means that either he accepts the particular means of technology and speaks their language, or he perishes."[2] Just the year before, in *Man and Technics*, Spengler warned, "This machine-technics will end with the Faustian civilization and one day will lie in fragments, forgotten – our railways and steamships as dead as the Roman roads and the Chinese wall, our giant cities and skyscrapers in ruins like old Memphis and Babylon."[3] For both men, technology was a "deal with the devil" that had put humanity under a powerful spell and was surely leading us all to disaster.

It is also around this time that Martin Heidegger began his analysis of technology. But, rather than criticizing or demonizing it, he attempted to uncover a deeper, hidden account of technology. For Heidegger, technology not only dominated contemporary life but also "enframed" nature and human civilization as a whole. The term "enframe" comes from the German word *gestell*, meaning frame, and indicates the way technology builds a new framework around existence, blocking the full expression of anything that falls outside of its boundaries. Everything becomes a product of technology or is "standing-reserve" – waiting to become a product of technology, just as the natural world, the human body, and human behaviour had become in the previous century.

Heidegger's concern about technology, as well as the terms enframing and standing-reserve, only began to appear in his work after the Second World War in the 1950s, but these ideas find their source much earlier in the 1920s, while he was teaching at Marburg University. Much of this period before Heidegger became widely celebrated for his groundbreaking philosophical work *Being and Time* was spent studying and lecturing on the work of Aristotle. Through a somewhat unconventional reading of the *Physics* and *Metaphysics*, Heidegger decided that our contemporary understanding of the natural world as a static heap of trees, dirt, and water was starkly different than the ancient conception. According to Aristotle, nature or *physis* instead indicates the overarching movement of all things or beings as they are born, grow, and die or come into being and go out of being. In various places, Heidegger explained this movement as unconcealment (a creative translation of the Greek word for "truth," *aletheia*) and concealment (from the Greek for forgetting, *lethe*), a disclosure and a hiding, or a presencing and an absencing. For Heidegger, an oak tree is unconcealed, disclosed, or presenced in the germination and growth of an acorn and concealed, hidden, or absenced once again as it falls and rots on the forest floor. Everything in one way or the other participates in this: trees, human beings, mountains, and massive stars as they form and die in deep space. In forgetting this movement, by viewing it as an inert thing, we were able to reconceptualize nature in the modern way Machiavelli did, transforming it into mere matter into which we could introduce whatever form we chose.

Heidegger was also convinced that originally *techne* was not opposed to this ancient understanding of nature. As touched upon in chapters 1 and 2, rather than coming into being through an internal efficient cause like an oak tree springing from an acorn, the external efficient cause of the craftsman "brings-forth" an artefact by working with the natural characteristics of its materials. And, as Heidegger agreed, the craftsman's unique role bringing into being things that would not be otherwise also characterized humanity's unique relationship with nature. Now, unlike ancient *techne*, Heidegger thought that technology does not "work with" nature but instead challenges it. So, the unconcealment of technology is characterized by the concealment of all other beings. Again, this overarching concealment or forgetting is what Heidegger calls enframing. Just as the hydroelectric dam on the Rhine conceals the movement of the river and submerges the river valley, turning it into a reservoir for power generation, technology as a whole obscures the entire planet, turning it into standing-reserve. The unconcealment of technology takes the diversified movement of nature and replaces it with the singular presence of technology.

All in all, Heidegger presented an extraordinary and upsetting vision of the world in the twentieth century. Everything, all of nature, was inexorably being incorporated, quantified and controlled, into the technological system. But, perhaps the most disturbing thing about Heidegger's vision was the idea that our capacity to understand that this was happening was itself being threatened by the final taking up of human beings as standing-reserve. By the 1960s, Heidegger began to even more urgently warn that the rise of technology was intrinsically linked to our inability to think about technology outside of its terms. In one of his last essays, the provocatively titled "The End of Philosophy and the Task of Thinking," Heidegger argued that the concealment of human beings by technology meant that we would no longer be able to think about or even notice the effects of the enframing process. "The need to ask about modern technology," Heidegger explained in un-characteristically succinct terms, "is presumably dying out to the same extent that technology more decisively characterizes and directs the appearance of the totality of the world and the position of man in it."[4] So, in the end, rather than the threat posed by any particular technology, whether nuclear annihilation or environmental catastrophe, it was our silence, our inability to think, act or question the crisis of technological envelopment that Heidegger pointed to as the supreme danger facing humanity. The real crisis, in other words, is that we did not and do not realize that we are in a crisis. As we sit on the edge of the vast abyss of technological nihilism, we remain completely oblivious to the danger.

And so we had now come to the critical point where the rise of technical knowledge almost completely dominated human behaviour and thinking. The centuries after Hobbes's articulation of the Great Reversal had seen the integration of technology into every aspect of our existence. And while we were clearly willing participants in the spread of technology as it delivered innumerable benefits and cures making our lives more pleasurable and less painful, Heidegger wanted us to realize that these products were only a corporeal manifestation of a much larger playing out of technology.

The Great Reversal Undone: A *Phronesis* Revival?

If the threat lies in the unlimited application of technical knowledge or the final ascendancy of kingly *techne*, then why not resurrect traditional limitations that had once kept technology in check? What about a *phronesis* revival?

While at Marburg, Heidegger considered the possibility of a return to *phronesis* as a way to respond to the quickly receding ability of human beings to intellectually understand and practically respond to the danger of technology. But he rejected the possibility of such a return. For the same reasons Heidegger concluded that existence is concealed by technology, he also decided that there remain insurmountable barriers to a revival of the practice of *phronesis*.[5]

Notably, he presented Aristotle's *phronesis* not as good judgment, practical reason, or prudence, but rather as "the right and proper way to be Dasein."[6] This strange "translation" reflected Heidegger's larger interest in not simply explaining the meaning of Aristotle's philosophical texts, but instead in engaging directly in the same ideas from his unique vantage point in the twentieth century. The German word

"Dasein," which literally means being-there (*da-sein*), is a core concept for Heidegger implying an authentic relationship with existence. For the Greeks, then, *phronesis* was the "right and proper way" to experience the true nature of the cosmos, existence, or Being. However, Heidegger thought that human beings in his day and age were being pushed in the opposite direction towards the inauthentic.[7] This inauthenticity is not simply a personal choice, but is reinforced by the social, political, and cultural institutions of modern technological society. A good example of our inauthenticity might be our refusal to embrace our essential finitude. Even though we may grudgingly accept the reality of our own mortality, our materialistic, youth-obsessed society pushes us to live our lives as though we should never age or are never going to die. In turn, we prioritize the accumulation of possessions and superficial good looks over our responsibilities to the next generation, whether through parenting, education, or as caretakers of the environment. It is as though our civilization insists that we ignore the fundamental contradiction at the heart of this inauthentic lifestyle. Therefore, in order to return to authenticity, Heidegger thought that we must either change the character of modern institutions or our relationship to them, or remove ourselves from them altogether.

The critical difference for Aristotle was that these institutional barriers "to the right and proper way" did not exist, at least not in the same all-encompassing manner. Instead, according to Heidegger, the virtue of *phronesis* allowed the ancient Greeks to constantly renew themselves, allowing them to overcome bad influences and bad habits that may have distracted them from their "right and proper" or authentic purpose and responsibilities. The "salvation of *phronesis*,"[8] he explained, allowed them to put aside the everyday things that kept them from living full and meaningful lives. So, if people became so preoccupied with sex that they were unable to think beyond their narrow desire or suffered from such severe despair that they were unable to think, *phronesis* gave them the capacity to break through their obsessions and depressions. As Heidegger put it, "Insofar as man himself is the object of the *aletheia* of *phronesis*, it must be characteristic of man that he is covered up to himself, does not see himself, such that he needs an explicit *a-letheia* in order to become transparent to himself."[9]

The unconcealment provided by *phronesis* was only required if there was first some need to clear some sort of obstacle. In other words, if the *phronimos* was never faced with personal, ethical, or political dilemmas, he would never be able to reveal himself as a person of good judgment.

For the Greeks and perhaps even for us today, how we react to the problems and challenges of everyday life tells our family, fellow citizens, and ourselves what kind of person we really are. Without these problems, there really is no opportunity to distinguish oneself as virtuous. In fact, despite the many problems we may still have, there seems a shrinking supply of these types of opportunities for virtue in a technological society. After all, the whole push of technology has been to solve problems. Perhaps that is why the bar has been lowered so that almost anyone might be considered virtuous or heroic for the most mundane of acts: parents taking care of their children, a patient recovering from a disease, a victim suffering from a crime, or an athlete hitting a home run. We feel the need to call these people heroes and their actions virtuous perhaps because we yearn for real heroes and real virtue where there may be very few and very little.

In a surprising way, it is the great success of technology that makes it easy for us to relinquish the practice of virtue and makes the return of *phronesis* more and more unlikely. Rather than taking on the difficult task of moderating the passions and overcoming bad habits, it is has become commonplace to seek out techniques, treatments, and medicines for what are now diagnosed as diseases and syndromes. We have in various ways attempted to treat the ups and downs of the human condition as simply a "condition" that needs some sort of cure. The development and practice of good judgment seems quaint if not impotent in comparison to powerful and effective technologies that function as external efficient causes on the human psyche.

Of course, for Heidegger, the point was that we cannot overcome the insurmountable barrier of technology in the same way the *phronimos* overcame the distracting pleasures and pains of everyday life. Where these things may have at one time been cleared away through good politics and ethics, we simply have no capacity to clear away technology.

The Good vs. The Authentic

For Aristotle, the presence of the *phronimos* assured that, even if the *polis* had lost its way, the bases of the good life could still be recovered.[10] However, by Heidegger's account, we had no *phronimos*, no repository for the good life to restore or return us to the right and proper way to be. Even if we wanted to return to the good life as the Greeks lived it, we would have to relearn all that had been forgotten with no teacher or exemplar to point us in the right direction. We, like archaeologists, would have to try to piece together a lost civilization from shattered artefacts.[11]

In turn, Heidegger's call for authenticity was really quite a different thing than Aristotle's discussion of the good life. Where a move to "the authentic" suggests the need to reject common standards, traditions, and institutions as fraudulent or corrupting, striving for "the good" requires a certain confidence in and loyalty to one's community and fellow citizens. Indeed, the *phronimos* is a pillar of the community, whereas lonely Dasein fights against convention and remains outside the system, dissenting, rebellious, and even seditious.

For example, his search for authenticity led Heidegger to praise the actions of the young saboteur Albert Leo Schlageter, who was killed in 1923 fighting against the French occupation of Germany during the interwar period and who was later lionized by the Nazis as an early martyr to their cause. He also endorsed early war mobilization efforts, commending men of "new courage" in his address as the first rector at Freiburg University after Hitler's election as chancellor in 1933.[12] He later explained that he thought that National Socialism would bring with it "a spiritual renewal" of the world, where "the same dreary technological frenzy, the same unrestricted organization of the average man" had caused the "spiritual decline of the earth" and drained "the last bit of spiritual energy that makes it possible to see the decline."[13] Heidegger's initial support of the Nazi party reflected his strong belief that the obstacle of modern technology had to be cleared away and that there was still a shrinking window of opportunity to do something to respond to the technological juggernaut that was crushing any and all possibility of a movement to a more authentic existence outside of the bounds of enframing.[14] The Nazis, Heidegger thought, would destroy the centres of modern technological society and replace them with a new spiritually rich culture. It was time for a different kind of radical politics that would sweep away the detritus of the old world and make room for something else. In turn, instead of *phronesis*, Heidegger appealed to a different kind of action he called "resolve."[15] He thought that resolve or resoluteness would open up or unlock a new route to the right and proper way, giving the German people a renewed ability to be authentic. Unlike *phronesis*, resolve required no understanding of the world that waited once these barriers were removed. This is what he meant when he said in his speech as rector of Freiburg University that Germans should stand "firm in the midst of the uncertainty of the totality of being."[16] The point here is that Heidegger did not want to make the mistake of positing a prototypical future or blueprint for utopia that would be built on the ruins of modern civilization. He did not want to restart the technological project with an updated version of a metaphysical good.[17] Instead, he demanded an acceptance of uncertainty or a total openness to whatever world would reveal itself from the rubble and ashes of what was to come. There was nothing practical or reasonable about Heideggerian resolve. It was creative destruction of the old order to make way for something new and as of yet unknown.

So, while *phronesis* and resolve are similar in their embrace of the spontaneous and unpredictable, the ancient virtue does not abandon certainty or the predictable altogether.[18] At the beginning of the *Ethics*, Aristotle asked, "Will not knowledge of [the chief good] have a great influence on life? Shall we not, like archers who have a mark to aim at, be more likely to hit upon what is right?"[19] The *phronimos* may have to be quick on his feet, able to adapt to changing circumstances on the ground, but he always keeps a general understanding of what these moves and modifications are aimed at. By contrast, Heidegger's resolute heroes were shooting blind. Because technology concealed or enframed their vision, they could not aim at their target but only remove what blocked it.

The Neo-Luddites

With the start of the Second World War, Heidegger gradually realized that his grand hopes for the Nazi movement were not going to be met.[20] While the matter is still a subject of great controversy, it is clear that he slunk away from his calls for resolve and began to consider different ways of properly responding to the challenge of technology. However, his earlier ideas are still reflected in much of today's anti-technology discourse. The "Neo-Luddite" movement, for example, seems to embrace similar tenets of Heideggerian resolve. Unlike their Luddite predecessors, who directed their ire at a particular technology, this collection of radical environmentalists and deep ecologists, groups like Earth First! and the vigilante Earth Liberation Front (ELF), the back-to-nature movement, the no-growth school, and strident anarcho-primitivists view the whole of technology as a danger and seek its quarantine, restriction, or total destruction. On a par with Heidegger's rejection of reigning institutions in the name of authenticity, these neo-Luddites seek alternative and often violent methods outside of contemporary politics and laws to achieve their goals.[21]

Ted Kaczynski is probably the best-known contemporary advocate of this kind of aggressive response to technology. In his manifesto *Industrial Society and Its Future*, he denies all possibility of reforming technology so that it would "prevent it from progressively narrowing our sphere of freedom." He instead calls for the complete overthrow of the whole technological system. Of course, Kaczynski's notoriety does not stem from the eloquence of his anti-technology theories, but from his twenty-year terrorist campaign as the Unabomber. Like Ernst Jünger and Oswald Spengler, the Unabomber envisioned a society that lay in ruins with only the strong and capable able to survive. But, rather than early-twentieth-century German idealists, Kaczynski was more likely inspired by fictional characters like George Hayduke, the protagonist from Edward Abbey's 1975 novel *The Monkey Wrench Gang*.[22] In this striking passage, Hayduke describes his vision of a post-technological world similar to Spengler's:

> When the cities are gone, he thought, and all the ruckus has died away, when sunflowers push up through the concrete and asphalt of the forgotten interstate freeways, when the Kremlin and the Pentagon are turned into nursing homes for generals, presidents and other such shitheads, when the glass-aluminum skyscraper tombs of Phoenix Arizona barely show above the sand dunes, why then, why then, why then by God maybe free men and wild women on horses, free women and wild men, can roam the sagebrush canyonlands in freedom – goddammit![23]

Perhaps Tyler Durden, the anti-hero of Chuck Palahniuk's 1996 novel *Fight Club*, will be inspiration for the next generation of neo-Luddites. In this speech, Durden describes a similar scene, left in the wake of the terrorist attack he dubbed "Project Mayhem": "Don't think of this as extinction. Think of this as downsizing … You'll hunt elk through the damp canyon forest around the ruins of Rockefeller Center, and dig clams next to the skeleton of the Space Needle leaning at a forty-five degree angle. We'll paint the skyscrapers with huge totem faces and goblin tikis, and every evening what's left of mankind will retreat to empty zoos and lock itself in cages as protection against bears and big cats and wolves that pace and watch us from outside the cage bars at night."[24]

These post-apocalyptic, post-holocaust scenarios are distinct from anything described by the eighteenth- and nineteenth-century critics of science and modernity. More anarchists than fascists, the neo-Luddites share the same vicious defiance of the Nazis and the violent atavism of the twenty-first-century Taliban and jihadist movements. Unlike George Orwell's character Winston Smith from *Nineteen Eighty-Four* or Aldous Huxley's Bernard Marx from *Brave New World*, Kaczynski and his fictional doppelgangers react to technological utopianism with visions of non-technological dystopia.

While crude and contradictory, the Unabomber's lengthy manifesto expresses many of the attitudes shared by more eloquent neo-Luddites such as Kirkpatrick Sale. Sale explains the common bond that links the movement together: "Wherever the neo-Luddites may be found, they are attempting to bear witness to the secret little truth that lies at the heart of the modern experience: Whatever its presumed benefits, of speed or ease or power or wealth, industrial *technology* comes at a price, and in the contemporary world that price is ever rising and ever threatening."[25]

So, whether it is the danger of mass hypnosis through television or mass sickness through industrial pollution, the neo-Luddites agree that technology is a threat to human and non-human life. In her short piece "Notes toward a Neo-Luddite Manifesto," Chellis Glendinning writes: "The worldview [the Luddites] supported was an older, more decentralized one espousing the interconnectedness of work, community, and family through craft guilds, village networks, and townships." She explains, "Like the early Luddites, we too are a desperate people seeking to protect the livelihoods, communities, and families we love, which lie on the verge of destruction."[26] Just as the introduction of new technology threatened the Luddites' way of life and community, Glendenning thinks that it also threatens ours. She goes onto to call for the dismantling of "destructive technologies" such as nuclear, chemical, genetic engineering, television, electromagnetic, and computer technologies. For her, these technologies serve as obstacles to the "life-enhancing worldview," whereas other technologies such as solar panels and wind power are somehow less problematic.

Authentic Technology

This highlights one of the more perplexing things about this faction of the contemporary anti-technology movement. The fantasy of returning to a primitive hunter-gatherer lifestyle or a decentralized medieval socio-economic structure would of course mean the abandonment of modern technology holus bolus, including solar panels, wind turbines, and hydrogen fuel cells. Obviously, these are no less, if not more, technological than any of the other technologies that Glendenning identifies as destructive. But, despite their various claims that they seek the obliteration of technology as a whole, every neo-Luddite has a long list of *particular* technologies that he or she love to hate, whether genetically modified organisms, internal combustion engines, or email.

Heidegger, however, did not really pick and choose which technologies to criticize or admit some were better than others. He thought that technology itself was enframing the planet and everything on it. He later clarified his position on whether it made sense to criticize particular technologies, writing: "For all of us, the arrangements, devices, and machinery of technology are to a greater or lesser extent indispensable. It would be foolish to attack technology blindly. It would be shortsighted to condemn it as the work of the devil."[27] He is even blunter in this interview: "I want to say that I am not against technology; I have never spoken against technology, nor against the so-called demonic elements in technology ... So, above all, the misunderstanding that I am against technology is to be rejected."[28]

There is no inconsistency here. After all, Heidegger's initial embrace of the Nazis was not the product of a naive hope that they would reject the technological as such. Obviously, he could never have had the expectation that the Nazis would fight the Russians and Americans with farmer's pitchforks and scythes.[29] Still, how can we reconcile Heidegger's call for resolute action against spiritual decline with these later statements? In his address as rector, Heidegger called for a recapturing of "the original Greek essence of science."[30] This recapturing entailed a move away from the contemporary effort to control nature through scientific research and technology and a return to the ancient Greek understanding of making or *techne*, which, he claimed, worked in cooperation or in partnership with nature.[31] As mentioned in chapter 1, he did not disparage the products of the ancient craftsman in the same way he did contemporary technology because they are "scenes of disclosure" for overpowering nature and draw our attention to the nature of existence.

This did not mean that Heidegger favoured a return to the simple, nostalgic world of the rural farmer of the Black Forest or the authentic ancient craftsman toiling away in his workshop.[32] Despite the fact that Heidegger often posed wooden bridges against hydroelectric dams and peasants farming against open-pit mining, the point for Heidegger is not "what" we build but rather "why" we build it. Indeed, the Pyramids and Parthenons of the ancient world were not quaint or modest projects by any estimate, and yet would still, for Heidegger, qualify as authentic artefacts. Therefore, Heidegger's post-technological world could still be grand and technically advanced in the same sense as these noble monuments. This seems to be what Heidegger is addressing when he proclaims that "the beginning exists still. It does not lie behind us as something long past, but it stands before us," it "has invaded our future; it stands there as the distant decree that orders us to recapture its greatness."[33] Rather than the aggression and violence associated with resolve, this moderate response does not require the destruction or restriction of technology, but instead a return to the building of authentic artefacts.

We get a further understanding of the possibility of authentic artefacts from Heidegger's discussion of a rural farm in his essay "Building Dwelling Thinking." He warns, "Our reference to the Black Forest farm in no way means that we should or could go back to building such houses; rather, it illustrates by a dwelling that *has been* how it was able to build."[34] Here he explicitly rejects a nostalgic return to some pre-technological age. In this same essay, Heidegger provides a remarkable analysis of a contemporary technology that seems to suggest the possibility of authentic technology: "The highway bridge is tied into the network of long-distance traffic, paced and calculated for maximum yield. Always and ever differently the bridge initiates the lingering and hastening ways of men to and fro ... The bridge *gathers*, as a passage that crosses, before the divinities – whether we explicitly think of, and visibly *give thanks for*, their presence, as in the figure of the saint of the bridge, or whether that divine presence is obstructed or even pushed wholly aside."[35]

Hubert Dreyfus and Charles Spinosa, two leading Heidegger scholars, explain that this unique passage shows Heidegger accepting that technological things such as highway bridges may allow for a "plurality of communities of focal celebration."[36] The modern highway bridge can open us up to a similar experience as the artefacts of the ancient world. At least in part, this answers what Dreyfus and Spinosa call "*the* question for our generation": "How can we relate ourselves to technology in a way that not only resists its devastation but also gives it a positive role in our lives?"[37]

The highway bridge can be taken as an example of Heidegger's claim that "we can affirm the unavoidable use of technical devices, and also deny them the right to dominate us, and so to warp, confuse and lay waste our nature."[38] While there is still an obvious antagonism here, the idea that we can "affirm" technical devices suggests that we can live with technologies while avoiding enframing and dehumanization. Then again, considering all that Heidegger has said, it remains unclear how we can live, work, and think in a technological society while not becoming dominated by technical devices.[39]

The Best Response Is No Response

As described above, Heidegger understood Nazism as a route to respond to the challenge of technology. And, while he moves away from this aggressive response, this should not be taken as an admission that it was non-viable. His unwillingness to explicitly disavow the goals of the National Socialist revolution suggests that he held out the faint hope that some time in the distant future a similar planetary effort to knock back and destroy the technological establishment would again be possible. In an oft-quoted interview given well after the war, he cryptically explains that the Nazis were "far too limited in their thinking" to fully realize or take advantage of the opportunity presented to them.[40] However, Heidegger does come to critique the Nazis because their revolution became a furthering of the "dreary technological frenzy" that led him to consider new ways to respond to the immediate challenge of technology.

At basis, the defeat of the Nazis brought Heidegger to question the very possibility of any contemporary political response to technology. He asks in a 1966 *Der Spiegel* interview, "How can a political system accommodate itself to the technological age, and which system would this be? ... We still have no way to respond to the essence of technology."[41] Heidegger is led to explore a far more passive approach. The recognition of the ineffectuality of a political or social response to technology is why he moves away from his call for a violent destruction of institutions to clear the way for authenticity and instead suggests that by accepting or realizing that technology dominates us we will once again know what it is to be in the grasp of a fate beyond our control. In other words, by realizing that technology is out of our control, we will move away from the technological impetuses to quantify and control.[42]

This call for passivity requires a stepping back from any and all activist effort to defeat or moderate technology. Heidegger comes to recognize that no politics and no programs of reform could themselves steer humanity away from the consequence of the challenge of technology: until we ourselves are taken up as standing-reserve we will not recognize the danger of our age. In *The Question Concerning Technology*, Heidegger writes: "The closer we come to the danger, the more brightly do the ways into the saving power begin to shine and the more questioning we become."[43] Only when we become fully cognizant of the supreme danger of technology will we be prepared to take a new course away from technological nihilism. Whether this will happen, what that course might be, and where it might take us remain a mystery.

We see similar calls for resignation and acceptance in many important twentieth-century thinkers. Lewis Mumford, for example, calls for "quiet acts of mental or physical withdrawal – in gestures of nonconformity, in abstentions, restrictions, inhibitions."[44] In *The Technological Society*, Jacques Ellul argues that we still have an opportunity to respond to the challenge of technology: "The challenge is not to scholars and university professors, but to all of us. At stake is our very life, and we shall need all the energy, inventiveness, imagination, goodness, and strength we can muster to triumph in our predicament."[45] But, like Heidegger, Ellul comes to give up this activism. In the later *The Technological System*, he questions whether it is at all possible for man to "'take in hand,' direct, organize, choose and orient technology,"[46] and decides, "Man in our society has no intellectual, moral, or spiritual reference point for judging and criticizing technology."[47] Marshall McLuhan has the same concern and, like Ellul, seems to straddle the moderate and the passive positions. In *The Gutenberg Galaxy*, he writes: "Far from being deterministic, however, the present study will, it is hoped, elucidate a principal factor in social change which may lead to a genuine increase of human autonomy."[48]

In both Ellul and McLuhan, there seems a small but quickly shrinking window of opportunity to do something to avoid or mitigate the onset of the technological system, or what McLuhan calls the global village. By contrast, George Grant argues that the window closed decades ago. Influenced by Heidegger, Grant contends that "the planetary technical future" is our "fate" and that there is nothing we can do about it.[49] He clearly states that "those who would try to divert, to limit, or even simply to stand in fear before some of [technology's] applications find themselves defenceless."[50] These thinkers decide that the activist effort to subordinate technology to human concerns is itself an outgrowth of technological thinking and actually seeds the way for further enframing. That is to say, protest and criticism of the "failures" of technology simply highlight the need for new methods to incorporate human needs into technology.

For Heidegger, passivity is simply another way for us to become open to the revealing of technology. By stepping back from the technological frenzy, we remove the primary obstacle to recognizing revealing. This is what he means when he quotes his favourite poet Hölderlin:

But where danger is, grows
The saving power also.[51]

So, even though technology is what threatens us the most, it is also the thing through which we might once again return to authenticity. When we come to realize through our own taking up as standing-reserve that we do not control the revealing of technology, but merely participate in that revealing, we may be able to return to a more authentic or free relationship with technology. Therefore, while passive, this approach is not also deterministic. In a sense, it is passivity with a purpose, helping to express the playing out of technology, whether or not it will over-whelm and conceal the essence of all other things, including ourselves.

Of course, we may find this approach frustrating: we should take to the streets, lobby for change, and take a proactive approach against the effects of technology. But, according to Heidegger, in order to escape all-encompassing technology, we must do nothing. Otherwise, our ac-tions will be sucked into the dynamo once more and turned out anew on the other side.

7   Virtue in a Technological Age

Chapter 1 began with scientist Bill Joy's warning about the hazards of genetic engineering, nanotechnology, and robotics. These new technologies, Joy proffered, represent a profound new threat to humanity. Similar warnings are not hard to find: rampant industrialization is destroying the environment; fossil fuels are dangerously warming the planet; genetically modified plants will lead to a catastrophic agricultural collapse; and so forth. While these threats are serious, they do not represent *fundamental* threats. In the past, the same sorts of concerns were raised about the printing press, the electric light, the radio, and of course the incredible risk posed by trains.[1]

Rather than fundamental threats, these technologies represent *transitory* threats. The continuing threat of pollution from car exhaust, for example, has been mitigated with the introduction of new technologies such as catalytic converters, higher-grade unleaded gasoline, and hybrid engines, all of which have reduced the toxins spewed into the air by individual motor vehicles. We have also been told by government and scientists that the introductions of electric engines and hydrogen fuel cells will further reduce if not eventually eliminate this particular threat. Consequently, we might conclude that air pollution and even global warming are big and worrisome but nonetheless transitory threats. Likewise, the grave problems related to the technologies that keep Joy up at night will likely be overcome through similar technical refinements or, simply put, the development of better technology. Joy's worries about misapplications, ineptitude, poor planning, and programming can all be addressed through the same process that has allowed for the cut in pollution emitted from individual cars. Of course, if we choose to do nothing, the accumulated effect of our inaction might well be disastrous, as Joy predicted.

But the *fundamental* threat of technology is not so easily solved. Instead of focusing on the inefficiencies or side-effects of particular technologies, the fundamental threat of technology is linked to our inability to think and act outside of its bounds. Heidegger argued that this inability is the result of the inexorable unconcealment of technology as it conceals all other things including human beings. However, the idea of the Great Reversal as it has been presented in the preceding pages attempts to demonstrate that this inability is not merely a consequence of a larger playing out of technology but the result of a set of choices. That is to say, the subordination of good judgment and the domination of technical knowledge is the outcome of a history of such choices.

To summarize, Augustine's early Christian rejection of good judgment stemmed from a distrust of worldly things, leading him to separate everyday decisions from the development of virtue and the good life. The introduction of a distinct "prudence of the flesh" and "prudence of the spirit" broke the vital connection between the lower goods associated with the body and the higher goods associated with spiritual satisfaction. Aquinas then turned our attention back to the everyday with a concept of natural law that proposed human society be reordered to match a divine plan hidden within nature. Reflecting Augustine's distrust of the lower goods, Aquinas argued that God had imbued humans with *synderesis*, or the rational capacity to recognize this divine plan and, with this knowledge, restructure human institutions to conform to it. This "top-down" approach is later echoed in Machiavelli's advice that all of nature is mere matter into which we can introduce whatever form we choose. Freed of the moderating influence of divine limitations, Machiavelli turned to the task of eliminating the influence of chance or *fortuna* by imposing human ends upon the world limited only by a lack of technical know-how. Hobbes subsequently argued that even this technical limitation could be overcome through time and industry, allowing for a total domination of the natural world and human nature. The remaining barrier to the achievement of this goal was the tenacious hold that fallible human judgments still had on social and political institutions. The Enlightenment thinkers next endeavoured to purge the influence of judgment and experiential knowledge once and for all from these institutions and empowered humanity to exploit nature towards true happiness and everlasting peace. This effort included the development of a social science that sought first to remake human society and then to remake the body and mind to rid them of any weakness, feeble-mindedness, and fallibility. In the twentieth century, there follows an infiltration of technology into all aspect of human life and society that brings with it both tremendous benefits and horrible destruction. From this point on there has been growing criticism of the invasive, dehumanizing character of contemporary technology and an earnest exploration of ways to limit or knock back the further envelopment of humanity into its fold.

And yet we still seem unable to stop its advance. Perhaps Heidegger's analysis of technology is so compelling because he provided an explanation of why we seem so impotent. He argued that, rather than being a tool of our own creation, technology is an autonomous being that actually uses humanity as a tool for its own expansion. Conversely, the argument here is that we are impotent not because technology is independent of our influence, but because somewhere along the line we began to relinquish the deeply rooted intellectual and practical capacities that allow us to understand and regulate the role of technology in our lives. In the modern world, we have consented to this constriction of human thought and action into the narrow confines of technical thinking in exchange for the satisfaction of our appetites and alleviation of our aversions. And, despite our growing reservations, we are still willing to make this exchange because we have lost our connection to a higher sense of purpose that once animated our search for the good life and happiness. Because technology seems to deliver everything that we need and want, we are unable to argue against its progress and helpless to prevent its further infiltration into the innermost recesses of our bodies and minds.

And yet, even in the wake of the Great Reversal, we may still be able to find and enforce limits on technology. Despite Heidegger's earlier rejection, this last chapter attempts to realize the third objective of this book by exploring the effort to revive the practice of *phronesis* and virtue in general as a response to the challenge of technology. This *"phronesis* revival" has become an increasingly important idea in contemporary scholarship and philosophy. Notably, the two leading advocates of the *phronesis* revival were among Heidegger's best students: Hannah Arendt and Hans-Georg Gadamer. They basically agreed with their teacher that technological domination represented the greatest threat to human life and society. But, rather than completely accepting his enframing argument, both argued that technology had come to this point of domination, not because of some playing out of Platonic metaphysics, but because we had given ourselves and our institutions over to it. In turn, they thought it still might be possible to extricate ourselves, our institutions, ethics, and politics from the cage of technology. However, this effort faces two major challenges.

First, the traditional foundation upon which the practice of *phronesis* used to be built has been considerably weakened. In turn, even if we somehow managed to get out from under the heavy weight of technology, we would have no clear idea how to act or think in a non-technological manner. So, even though Arendt and Gadamer argued that the ancient virtue should be returned to its rightful place of priority above *techne*, they offered no obvious starting place for humanity to relearn its practice. Nonetheless, both thinkers, who studied at Marburg with Heidegger in the 1920s, proposed that there were still opportunities for action and thought that stood outside of the demands of technical thinking. They offered *phronesis* as an alternative because it is based on the central unpredictability of opinions, people, conditions, and circumstances. It allows for the introduction of individual experience and a community-based ethics and politics thought to stand outside of the rigidity of *techne*. Yet, they also agreed with Heidegger that in the twentieth century our society had been inculcated by technology, with very few areas left untrodden. Our world differs radically from Aristotle's because human thought and action are clouded by scientific and technical expertise. This is a significant difference because, if we are truly suffering under the hegemonic influence of science and technology, as Arendt and Gadamer have described, we cannot be expected to have enough unfettered judgment, good habits, and virtue left to make right and proper choices once the tyranny of technological domination has been dismissed or removed. Simply dethroning technology from its dominant role will not mean that the individual citizen will all of a sudden become a good and responsible decision maker. Remember, in Aristotle's description, that *phronesis* was the fount of establishment thinking and the *phronimos* required a large support network of good parents, friends, teachers, and political leaders. But according to both of these twentieth-century thinkers, this foundation of education and role models is the very thing missing in contemporary society. Arguably, in a world dominated by science and technology such as ours, there are no persons to pass down the lessons and practical experience required for the practice of the virtue. Gadamer, for instance, vigorously argued that technology defines the thoughts and actions of the average citizen, sets new terms for human existence, manipulates our minds, and even denies our identities.[2] This suggests that in a society dominated by technology there is no ground upon which the *phronimos* can stand. Technology has supplanted the very conditions through which practice, experience, understanding, and interpretation can inform good judgment.

This is the first basic challenge to be faced with any effort to revive *phronesis* in a technological age. We have handed over our decision-making procedures to a range of technical experts, specialists, and managers and have thus left few if any sources for relearning the practice of the virtue. Our politics, our laws, our educational institutions, and even our communities are planned and managed by technical experts. Very few areas of life and society remain unaffected by or immune to their influence. This observation might lead us to find common purpose in the response of the neo-Luddites discussed in the previous chapter. By purging technology from our society or by withdrawing to the remaining non-technological corners of the planet, we might hope that the practice of virtue might one day re-emerge. However, the goal of the *phronesis* revival is to learn to live with technology, to find and enforce limits upon it, rather than to escape from it.

This leads to the second challenge of any *phronesis* revival as well as the second objective of the book: how in a technically run *polis* human beings become ever more worked upon as mere material and how this kind of treatment may actually bar the practice of *phronesis*. Remember Aristotle's straightforward warning about the products of *techne* from the *Politics*: "It is for the sake of the soul that these other things [external goods] are desirable, and should accordingly be desired by every man of good sense – not the soul for the sake of them."[3] This means that technical innovation must be directed by the higher virtues such as those associated with family, community, education, politics, and philosophy. It needs to be pointed out again that there is no disagreement from Aristotle that human beings need the products of *techne*. Not only is it senseless to suffer from cold, hunger, and pain, but these basic deprivations bar us from attaining deeper happiness. The virtue of *techne* is that it clears the way of these sorts of distractions and, therefore, participates in the achievement of higher human goods – the alleviation of basic bodily needs is "for the sake of" the higher things of the soul – good, beautiful, and noble things.

Aristotle's discussion of virtue asserts that technical production has to be preceded by the ethical mastery or self-discipline of the passions. As he says, only a man of "good sense" should desire "these other things." In other words, if the bodily needs and appetites that inspire technical production remain without virtue, undisciplined, defective, or excessive, then the products of *techne* will not be in proportion with the right and proper needs of the citizenry. In fact, a man of "bad sense" will necessarily desire products detrimental to himself and his community. Fortunately, even in such an instance, if his community has "good sense" on the whole, then good laws and other regulations will limit the vicious desire of this man. That is to say, he will not be given the opportunity to articulate his bad sense into harmful products. Society will not allow Dr Frankenstein to animate his monster.

Only when our judgments about higher things subordinate *techne* do we need not worry about quarantining areas of our lives off from technology or have to run to the hills, as is the advice of the neo-Luddites. That which is most essential to being human will remain unencumbered, undetermined, or unbound by mere products because those products will be determined by those essential things. Research and development in medicine, for example, will be determined by our want for happiness rather than our happiness being determined by medicine. When this order is upset and the lower takes precedence over the higher, technical production comes to dominate our lives and our thinking.

And, at present, this order has indeed been upset. Rather than learning the classical lesson on developing an internal ethical mastery of the passions, we have largely accepted the idea that we require an external efficient cause to both regulate and fulfil our desires. The "Artificiall Chains" that restricted an "excess of passions" described by Hobbes in the *Leviathan* now take the form of remedies and cures for what are viewed as a set of various chemical, metabolic, and biological imbalances and deficiencies. While there is little doubt that these treatments alleviate pain and suffering of all kinds, their effectiveness and ubiquity make it all the more unlikely that we will be able or willing to choose Aristotle's path to virtue.

Virtutropics

With the "bad sense" we have derived from our narrow technical outlook, we have been compelled to create a set of "virtutropic" (virtue + *trope*) technologies, to coin a phrase, designed as substitutes for the traditional ethical virtues. The Greek word *trope* means turn, bend, or change and is used as a suffix in words such heliotropic, referring to the way plants turn or bend towards the sun, and psychotropic, a class of drugs designed to change thought, mood, or behaviour. So, a virtutropic technology refers to a technology that turns, redirects, or changes the practice of virtue or excellence.

To be clear, the four examples of virtutropics listed below have already been subject to broad criticism: antidepressants have been reproached for being overhyped in their effectiveness in fighting depression; steroids for the array of negative physical ailments resulting from their overuse; elective cosmetic plastic surgery as a drain on limited medical resources; and gastric bypass surgery as unnecessarily risking the health of patients. However, the problem with these types of criticisms is that they focus on the transitory threat posed by these interventions. If their benefits could be boosted and side-effects ameliorated, if the technology could be improved, would critics then become advocates for the wider prescription of these drugs and performance of these surgeries? At the very least, they would be left with no evidence to support their concerns. The idea behind the list below is to explore the fundamental threat posed by virtutropics, not to list their technical deficiencies. Really, it is the fact these technologies are so global in their effect and so discreet, becoming less and less conspicuous to both the user and those around them, that makes them so hazardous.

Pharmacological Virtutropics

Psychopharmaceuticals

The use of psychopharmaceuticals to treat mental disease has become more and more common since the first generation of antidepressants was introduced in the 1950s. But in 1987 there was an unprecedented acceleration in their use when fluoxetine, commercially available as Prozac and marketed as one of the first selective serotonin reuptake inhibitors (SSRI), was approved for prescription in the United States. Not long thereafter, the drug became as much a cultural phenomenon as a medical breakthrough, ringing in a new era of "cosmetic pharmacology" that saw its widespread use not just for psychiatric treatment but also for self-improvement. Peter Kramer, who introduced the term "cosmetic pharmacology" in his 1993 book *Listening to Prozac*, was an early critic. He worried that, because it had such a tremendous influence on its users, Prozac might instigate a wider societal embrace of the drug for everyday emotional and intellectual augmentation. He writes:

"It is all very well for drugs to do small things: to induce sleep, to allay anxiety, to ameliorate a well-recognized syndrome. But for a drug's effect to be so global – to extend to social popularity, business acumen, self-image, energy, flexibility, sexual appeal – touches too closely on fantasies about medication for the mind."[4]

According to Kramer, in contrast to the overprescription of tranquillizers such as Valium (the most prescribed drug in the United States from 1969 to 1982) a generation ago, Prozac was seen not simply as a temporary respite from everyday drudgery, a source of relief, or means of treatment but also as a personality enhancer. People began taking it to "get ahead" not just to "get by." After all, the drug seemed an easy alternative to traditional avenues to personal success. Rather than relying on a healthy family life or good education to nurture positive character traits, individuals could now turn to an enthusiastically prescribed pill. Unlike other drugs, Prozac's "global" effectiveness presaged a time when we could replace or circumvent the antiquated and time-consuming teaching of the virtues to our children and students with a custom-made chemical infusion.

The problem with this circumvention is that it skips over a traditional and critical step of personal development. Considering the observations of one social worker, Kramer contends: "The medication had done what she would have wished to accomplish with psychotherapy: it had facilitated an improvement in the family dynamics. The problem, for the social worker, was that this change came about without any increased self-knowledge on Julia's part ... She believed that medication-induced change, unaccompanied by growth in self-understanding, was inferior to what psychotherapy has to offer."[5]

While it would be unwise to draw general conclusions from one case, it is fair to say that the drug treatment is successful in changing the behaviour of most patients, including "Julia." But, instead of her good behaviour coming from Julia herself, Prozac is the real source. Her actions do not reflect her upbringing, community, individual character, or "self-knowledge," but the design of the drug. Thus, she has not actually learned what it takes to act properly, and yet she nonetheless appears to act in a proper way.

Before continuing, it needs to be asked, if a drug works so well and provides such a positive result, what does it matter? Indeed, someone who is suffering from pain will likely care little about the source of their relief or whether they have missed out on an ethical education as described by the ancient Greek philosopher Aristotle. And, in all likelihood, it is true that individual or isolated cases do not in and of themselves represent any kind of threat to humanity. However, the widespread use of these drugs suggests something far more disturbing.

The problem of circumvention, the skipping over the ethical mastery of the passions, could trigger a larger societal effect that goes well beyond immediate users. Following the logic of Aristotle, virtuous action is first learned through imitation of a *phronimos* or person of good judgment. As it was put at the start of chapter 3, the *phronimos* is not simply lucky or blessed to live a good life, but has deliberately, knowingly, and meritoriously chosen a path that leads to happiness. The *phronimos* appear to be happy because he or she has successfully ordered their lives based on experience and a general knowledge of the way the world works. In turn, their fellow citizens can easily identify them as a good role model and be confident that in imitating their actions they will follow a similar path to happiness.

But, in the case of someone on Prozac there is an appearance of happiness that is not backed up with the aforementioned prerequisites of experience and knowledge. The change in Julia, as noted above, came without the hard work associated with rethinking and reordering the way she lived. Consequently, her social worker worried that if she was taken off the drug the very same patterns and problems that led her into depression would simply resurface. It follows that, even though she may be happy and thus appear to be a good role model, imitating her actions will not lead anyone down a path to happiness.

With anywhere between 5 and 10 per cent of the general population of the United States on these kinds of medications and trends suggesting a further substantial increase year by year,[6] this scenario will become more common and problematic. As the application of these drugs broadens, we may no longer be able look to the most happy, courageous, or intelligent among us as role models for living a good life. Next-generation antidepressants (e.g., new triple reuptake inhibitors or TRIs) will even more successfully produce happiness; new anti-anxiety drugs will kindle a chemically induced bravery; and newly developed nootropic (taken from the intellectual virtue *nous* + *trope*) drugs will soon provide an easily accessible external enhancement for learning acquisition.[7]

Rather than passing down an *ethos* for the practice of the good life, we will leave the next generation with little alternative but to take these drugs as well. This has already started with the prescription of behaviour-modifying drugs to children. Ritalin and other stimulants such as Adderall are designed to influence a child's biochemistry, allowing them to concentrate in the classroom.[8] Compared to the virtually instantaneous results achieved by these drugs, traditional methods of education seem inefficient. But for a child to take these drugs at an early age means that they will have less opportunity to learn to master their emotions or gain self-discipline and self-understanding. Yet, is not the point of education to teach children things such as concentration, patience, and moderation? To simply impose these attributes on a child in pill form is, to say the least, contradictory to these ends.

Steroids and Performance-Enhancing Drugs

While this description of psychopharmaceuticals as a virtutropic technology may seem unfamiliar, the same basic scenario has already been demonstrated in the case of anabolic-androgenic steroid use. From the Greek words *anabole* (building up) and *andro-* (of a man) plus *-genic* (producing), this class of steroids has been taken as a performance enhancer by athletes, such as weightlifters and bodybuilders, since the 1950s to build bigger muscles and to assist in quicker recovery from training stress and injury. Later, the use of anabolic steroids became common among Olympians before the drugs were banned from competition in the mid-1970s. In the early 1980s, steroid use increased among professional athletes and also infiltrated college and high-school sports.[9] Prohibited more broadly with tougher legislation in the late 1980s and early 1990s, the non-medical use of anabolic steroids is now restricted across much of the world.

These bans are largely based on the array of negative health effects caused by prolonged use of these drugs. However, the first "International Olympic Charter Against Doping in Sport" also listed other reasons: "The use of doping agents in sports is both unhealthy and contrary to the ethics of sport, and that it is necessary to protect the physical and spiritual health of athletes, the values of fair play and of competition, the integrity and unity of sport, and the rights of those who take part in it at whatever level."

Along with the health of athletes, there is also concern for the ethics of sport, fair play, integrity, teamwork, and the rights of participants. In other words, because doping provides an unfair advantage to its users, it amounts to cheating. Nonetheless, because they provide such a tremendous competitive edge, many athletes continue to take these banned substances. Well after the prohibitions were put into place, scores of high-profile sprinters, cyclists, professional football and baseball players, and so on were either caught with the drugs in their system or admitted to using them. Despite ever more rigorous and sophisticated testing methods, it is generally accepted among the members of the sporting community that the chemists who design new steroids seem able to stay one step ahead of the anti-doping agencies.[10]

In the aftermath of the Major League Baseball doping scandal of the last decade, many people expressed concern about the effect it would have upon younger athletes. One journalist explained the obvious logic in his article "Young players need a few good role models":

> Major league trends have always trickled down through the minors, colleges, high schools and even into the little leagues. Until they saw it on TV, how did Little League World Series champs know to run out of the dugout and pile on top of each other? When did college pitchers start headhunting? How did high school batters learn to charge the mound?
>
> Their heroes did it …
>
> The pros need to know the public has no tolerance for cheaters.
>
> Youngsters need to know that it's not worth risking their health just to imitate the guys on TV.
>
> A tough and airtight steroid policy can be a watershed moment in the game's history. If the players listen and follow, the rest of the nation will be right behind them.[11]

The worry is that no matter how naturally gifted a promising athlete may be, he will simply be unable to achieve the same results as his role model unless he also take steroids or some other performance-enhancing drug (PED). Trying to emulate their chemically enhanced heroes, children will skip over the critical step of learning the discipline and work ethic traditionally associated with high-performance athleticism.

Parents and coaches worry about the tremendous health risks. But even more problematic is the possibility of next-generation drugs offering performance enhancement with a far smaller chance of negative side-effects. Newly developed non-steroidal selective androgen receptor modulators, or SARMs, for example, promise to provide the same results as anabolic steroids but with notably less danger of organ damage.[12] With further advancements, it seems likely that athletes will be able to take performance-enhancing drugs without any risk to their health, undercutting the original reason for the ban. In turn, it is not difficult to envision a time when the prohibition on PEDs could be lifted, allowing athletes to employ them as just another one of their many training aids. That is to say, they would be part of the same spirit of fair play that already allows well-funded athletes access to better equipment, superior coaching, and other advanced training techniques not available to their equally talented opponents. In turn, this practice will "trickle down" to the next generation of competitors.

The problem again is that the criticisms of steroids and PEDs have focused on the transitory threat they pose. Once technical deficiencies are reduced or eliminated, these arguments will have little sway. With the health risk gone, young athletes will have little reason or incentive not to take them. Rather than passing down an ethic of good sportsmanship and hard work, we will leave the next generation with little alternative but to take these drugs if they ever hope to compete against their enhanced opponents. As a result, even if we wanted to return to the practice of traditional sport, we will have no exemplars to lead the way.

Surgical Virtutropics

Plastic Surgery

Another familiar example of a virtutropic technology is plastic surgery. The practice of body modification has a long and diverse history. Almost every body area, from foot binding in ancient China to cranial deformation in seventeenth-century France, has been subject to some form of artificial manipulation. Practices such as tattooing and piercing remain common examples of body modification today. While often painful and sometimes cruel, these practices are steeped in the customs and traditions of particular cultures. By contrast, plastic surgery permits an almost indefinite and continual modification of physical features.

Modern reconstructive plastic surgery finds its origins in experimental operations to correct major facial injuries suffered by soldiers in the First World War and birth defects in children such as cleft palates. Like cosmetic pharmacology, the field of cosmetic plastic surgery has developed as these procedures are applied to non-therapeutic, elective enhancements. Most often, patients seek to enhance the appearance of youth, fitness, or some other idea of attractiveness. But in changing their physical appearance, cosmetic surgery patients also seek some sort of psychological or inner change. Plenitas, a successful plastic surgery clinic in Argentina, posted this typical description of a patient's motivation for the surgery on their website:

"I want to gain self-esteem and feel comfortable with myself." – Michelle

Michelle is 35 years old, she's a gym trainer and lives in Panama City, Florida, U.S.A. Being a mother of four children, she combines her family life with her work as a fitness instructor.

As a consequence of her two pregnancies, Michelle's body changed completely, which led her to consider the possibility of having plastic surgery abroad. Seeing herself young and going back to her size would allow her to encourage her students to train harder, a necessary step to meet her own goals.

Michelle chose us because she believed that Plenitas would contribute to both improving her physical appearance and changing her inner self.[13]

Of course, the kind of confidence that Michelle seeks is usually realized through a sense of accomplishment after exercise and training. Certainly, a fitness instructor with a lean and strong body gains the respect of her students and colleagues because it is an obvious indication that she is good at what she does. As with the example of PEDs, undergoing a liposuction procedure merely gives the appearance of accomplishment and knowledge.

Obviously, cosmetic surgery is in good part an attempt to fool or trick others into thinking that there is a real connection between one's post-operation appearance and a healthy lifestyle or background. We are misled to believe that because these individuals have the good insight, confidence, and moderation to properly satisfy the lower goods associated with the body, they also are examples and role models for everyone. But, as the saying goes, appearances can be deceiving. As is the case with Julia, any effort to imitate Michelle's lifestyle under the belief that it would deliver us similar results would be futile.

Gastric-Bypass Surgery

Where liposuction is cosmetic, the gastric bypass is designed to permanently bend the will towards restraint. Stomach stapling, basically decreasing the size of the stomach and thus allowing only a small intake of food at one time, has become a far more common and safer procedure in the last few years. Doctors are recommending the surgery to help patients control diabetes, heart disease, high blood pressure, degrading joints, and other ailments associated with morbid obesity.[14] And, because rates of obesity have skyrocketed across the world, this surgery is in very high demand.

The problem of circumvention associated with this surgery applies in a slightly different way than with the virtutropic technologies presented above. A person who has undergone this procedure will lose weight because they are forced to starve themselves. Because this may lead to malnutrition and related illnesses such as osteoporosis and even beriberi,[15] patients must commit to a regime of vitamin and nutritional supplements to maintain their health. All told, there seems to be a swing from a vicious excess to a vicious deficiency, leaving out the virtue of moderation altogether. But, for the morbidly obese, it may be that some sort of psychological or physiological trait does not allow them to regulate their food intake and practise the virtue of moderation by themselves. By literally bypassing or circumventing the physical source of their immoderation (i.e., their stomachs), these individuals hope they can live longer, better, happier lives.

Yet again, the problem is that the individuals who undergo this surgery have not learned to be moderate or gained an ethical mastery of their passions, yet it appears to others by their weight loss or skinny physique that they have. As with the other examples presented above, rather than being an exhibition of self-control and a healthy lifestyle, the will of gastric-bypass recipients is bent by some external cause. In turn, they appear to exercise moderation, are mistakenly recognized as virtuous, and might be considered role models when in truth and despite all appearances they are not.

Admittedly, this short list represents just a preliminary sketch of the new problem of virtutropic technologies. Amazing innovations in prosthetic limbs, ears, eyes, and other mechanical enhancements to the human body and senses; the imminent arrival of discreet human/computer interfaces; and the potentially endless alterations that will be made possible by genetic intervention and transgenerational modification suggests this problem will become more common and acute. At least as they have been presented above, these technologies offer clear alternatives to traditional routes to achieving a self-mastery of the passions or developing an internal efficient cause towards virtue. Rather than learning the virtues of good judgment, fortitude, strength, and moderation, we instead turn to technologies that provide us with the appearance of these virtues. With outmost sincerity and sympathy, the idea here is not to cantankerously scold individuals who use these technologies or to point out some weakness in their individual character. As noted earlier, no one can really be expected to turn down the promise of a therapy or cure for chronic pain, whether it is physiological, emotional, or psychological. The real concern has little to do with individual cases. Instead, the idea is to highlight a larger ethical and political impact of virtutropics that goes well beyond the individual user. As these technologies become more ubiquitous in use and discreet in function, future generations will be faced with a remarkable and novel dilemma. Lacking a strong foundation for the learning of virtue and seeing no dependable role models for its practice, citizens of the future will have little choice but to accept virtutropics as their one and only route to health and happiness. Thus, it seems prudent to regulate or prohibit virtutropic technologies because of the fundamental risk they pose. For similar reasons that reproductive cloning was banned despite the tremendous curative benefits it offers, the use of virtutropics should be limited.

Transhumanism

Of course, this whole discussion on the great importance of virtue may still seem quite foreign if not unconvincing. After all, is not life just bodily needs, appetites, and desires? Are not humans made up of nothing more than the waxing and waning of various pleasures and pains? More and more, neuroscientists tell us that we are composed of a complex network of electrical impulses, transmitters, and receptors awash in a sea of rising and falling brain chemicals. By providing convincing and concrete physical explanations for what were formerly thought to be mysterious vagaries of human action and thought, neuroscience seems to verify that we are nothing but "matter in motion," just as Hobbes theorized centuries ago. And, if this is the case, the technical knowledge associated with "lower goods" is, in fact, the only kind of knowledge we can have. By this account, all we really know is self-preservation. Our bodies and our minds are nothing more than "human material" to be manipulated towards desired ends. If all we are is matter or material, then it is our most basic right to have access to any technology that contributes to our further survival, however perceived. It really does not matter if it is by internal or external mastery, through self-control or technological intervention. We simply need not worry about "higher goods" at all.

It is from this perhaps more familiar perspective that the contemporary transhumanist movement argues that we should use and develop ever newer technology to improve our intellectual and physical capabilities and purposefully evolve or transcend the human mind and body in their traditional forms. George Dvorsky, a leading transhumanist and member of the board of directors of Humanity+, the new moniker of the World Transhumanist Association, notes that "Transhumanists recognize that their bodies are a kind of machine," and sees that

> a growing number of people are turning to transhumanism, which aims to promote and encourage human enhancement through the application of science and technology. With roots in humanist and Enlightenment thinking, transhumanism is an emerging broadly based philosophy, bioethics, cultural phenomenon and lifestyle choice whose proponents believe that technology can and should be applied to improve the human condition. Transhumanists believe that humanity ought to enter into a post-Darwinian phase of existence where intelligences, rather than the blind forces of natural selection, are in control of their own evolution.[16]

And so, guided by the very same values as modernity to extend, improve, and affirm human life, the transhumanists seek to upgrade the human machine through whatever mechanical, computer, chemical, or genetic intervention is necessary. The modern term "transhumanism" is thought to have been used for the first time in a 1927 book by Julian Huxley, the noted evolutionary biologist, past president of the British Eugenics Society, and brother of *Doors of Perception and Heaven and Hell* author Aldous Huxley.[17] Later, Huxley reflected further on the great progress humanity has made over the last one hundred years, observing that "up till now human life has generally been, as Hobbes described it, 'nasty, brutish and short'; the great majority of human beings (if they have not already died young) have been afflicted with misery in one form or another – poverty, disease, ill-health, over-work, cruelty, or oppression."

Huxley argued that science and technology should be put in service to lighten our misery:

> It is as if man had been suddenly appointed managing director of the biggest business of all, the business of evolution – appointed without being asked if he wanted it, and without proper warning and preparation. What is more, he can't refuse the job. Whether he wants to or not, whether he is conscious of what he is doing or not, he is in point of fact determining the future direction of evolution on this earth. That is his inescapable destiny, and the sooner he realizes it and starts believing in it, the better for all concerned.

He concluded that "the exploration of human nature and its possibilities has scarcely begun. A vast New World of uncharted possibilities awaits its Columbus."[18] The remarkable thing about Dvorsky, Huxley, and other advocates of transhumanism is their embrace of an astoundingly lucid technological vision of the future that seems almost completely unencumbered by older sentiments that may still persist long after the Great Reversal. And while they share the same basic goals of Francis Galton and the earlier eugenicist movement, their means are clearly different and open to individual "lifestyle choice." For them, we can all become our own craftsman-king, able to quantify and control our bodies and minds in any which way we choose.

Critics call the transhumanists everything from hubristic to genetic elitists. Francis Fukuyama even worries that enhanced individuals may eventually abandon their democratic tone and claim that their intellectual and physical superiority also affords them more rights and privileges than unenhanced citizens.[19] But the real danger of transhumanism is its potential for technological relativism and, in turn, technological nihilism. The point here is that, even though the transhumanists may seem on the fringes or to have read too much science fiction as children, their idea is a logical consequence of the narrowing of human thought and action to a singular technological vision of the future of human life and society. And, rather than science fiction, we already have the basic technological infrastructure in place to empower this vision. Without enforceable limits on the transhuman project and under the pretext of therapy, individuals will evolve/enhance themselves in an indefinite variety of ways depending upon their individual needs and desires. In stages, the unlimited enhancement of the human body and mind may eventually splinter or fragment humanity into a wide array of new technologically evolved species. As conceptions of happiness and the requirements of health diverge, as the senses are recalibrated, what might have begun as a common goal to increase pleasure and reduce pain ends up creating wholly new and tremendously diverse transhuman ambitions. If these alterations become widespread and intergenerational, spinning off into the invention of new bodily forms and new ways of thinking delinked from anything recognizable, the traditional idea of humanity will fade away.

The Lost City

And while it is fair to say that the vast majority of people do not adhere to the transhumanist idea, the hard task remains to articulate the reason for our hesitancy. One response to both the claims of neuroscience and transhumanism is that we cannot fully understand nor should we radically alter what is human because it was created as part of a divine plan. However, the "sacred boundaries" argument has little sway over non-believers. In fact, there has been a strong counter-response by the scientific establishment contending that belief, religion, and God are all consequences of biology and brain chemistry. Titles such as *Why We Believe What We Believe: Uncovering Our Biological Need for Meaning, Spirituality, and Truth*; *Why God Won't Go Away: Brain Science and the Biology of Belief*; and *The "God" Part of the Brain: A Scientific Interpretation of Human Spirituality and God* assert that advanced brain scanning techniques can prove once and for all that religious belief is simply all in our minds. These scientists argue that they can easily replicate the "religious experience" by injecting patients with dopamine or by having them ingest hallucinogenic drugs such as psilocybin. In one experiment, scans of these drugged patients were placed side by side scans of Buddhist monks meditating. This comparison showed that both drugs and prayer affect the same part of the brain (i.e., the medial temporal lobe) and induce the same overall state of being. Rather than being closer to God or at one with the cosmos, these monks have simply learned how to trigger certain neural receptors.

This same group of scientists understands that happiness is really a similar trick of the mind. And yet, while the brain of a person on a psychopharmaceutical or some other enhancement may light up in the same way as a happy person not on the drug, these scans cannot show whether this state was induced by internal or external mastery of the passions. The critical difference between these two roads is that someone travelling down the road to internal mastery has learned through practice and experience how to live in harmony with their everyday circumstances and is able to guide those around them to a similar harmony. Of course, to ask everyone to simply master their passions with no assistance from "external goods" is not only cruel but also unrealistic. Just like the citizens of the "first city" mentioned at the start of this book, we cannot expect nor do we want to sustain ourselves without technology.

But the road to external mastery is more problematic. This road leads us to a strange and troubling dissonance between personal happiness and the well-being of our communities. Rather than a harmony attained between community and self, the destination at the end of this road is where the transhumanists are heading: to develop better drugs and other enhancements, to advance the technological infrastructure, to better quantify and control the functions of both the body and the mind. This road brings us to the "second city" where human beings are treated as mere material to be manipulated and formed in any which way. Critically, the real threat of living in the *techno*-polis is not the manipulation of the human form as such but the inability to find and enforce limits on that manipulation. Without an ethical compass to guide it and the political will to rule it, technology will in time replace the internal efficient cause of human beings with an external efficient cause. In turn, all of nature, both human and non-human, will rely on technology for its birth and growth. We see this already in the development of genetically modified crops and organisms.

A more fruitful response to the challenge posed by transhumanism and the problem of technology in general is to refound the older, unified lost city, re-establishing the earlier hierarchy between *techne* and *phronesis*. Rather than the *technites*, this city is led by the *phronimos*, who puts a priority on family, education, community, and law. These things are not put in a place of priority out of some sanctimonious moral duty or conservative ideology, but because they provide us with a foundation to make decisions about how to live a good life. The dynamic relationship or complex interplay between parents and children, teachers and students, friends and citizens, and among human beings in general is the basis upon which we gain a rational understanding of what we need and learn the habit of good judgment to deliver it. This city gives us an irreplaceable "bottom-up" insight into the needs of our bodies and minds, emotions, and psyches that, if satisfied, leads us to a sense of fulfilment and happiness. It also reinforces the need to put technology in the service of these ends rather than putting these ends in the service of technology.

To strengthen the considerably weakened foundations upon which we might relearn the lost virtue of *phronesis* will require a significant rethinking of the way we build and run our educational, social, and political institutions. Thus, this work of theory ends with an already quoted passage from Hobbes: "I recover some hope, that one time or other, this writing of mine, may fall into the hands of a Sovereign, who will ... convert this Truth of Speculation, into the Utility of Practice."[20]

Notes

Introduction

1 William Wordsworth laments:

Our meddling intellect
Misshapes the beauteous forms of things: –
We murder to dissect.

"The Tables Turned," in *Lyrical Ballads, with a Few Other Poems*, originally published in 1798 with Samuel Coleridge (London: J. & A. Arch, Gracechurch Street).

William Blake soon after bewails "these dark Satanic mills" that have sprung up across "England's green & pleasant Land" ("Milton," in *The Complete Poetry and Prose of William Blake*, ed. David V. Erdman [Berkeley: University of California Press, 1982]). Paul Cantor notes in his essay "Romanticism and Technology: Satanic Verses and Satanic Mills" that "though some scholars have questioned whether Blake had industrial mills specifically in mind here, in the common understanding of these verses Blake views the so-called technological progress in England as an act of desecration" (*Technology in the Western Political Tradition*, ed. Arthur Melzer, Jerry Weinberger, and M. Richard Zinman [Ithaca: Cornell University Press, 1993], 110).

Henry David Thoreau focuses his ire on the new railroad system that is spreading like a web across the American countryside: "That devilish Iron Horse, whose ear-rending neigh is heard through the tow, has muddied the

Boiling Spring with his foot, and he it is that has browsed off all the woods on Walden shore ...," deciding "We do not ride on the railroad; it rides upon us" (*Walden and Civil Disobedience* [New York: Penguin, 1986], 136.

2 For the sake of readability, all ancient Greek terms have been transliterated and presented without diacritical marks or accents.

3 In *Beyond Objectivism and Relativism: Science, Hermeneutics, and Praxis* (Philadelphia: University of Pennsylvania Press, 1983), Richard Bernstein identifies *phronesis* as the "underlying common vision" of some of the most important and influential thinkers of our time including Hans-Georg Gadamer, Jürgen Habermas, Richard Rorty, and Hannah Arendt. Ronald Beiner's *Political Judgment* (Chicago: University of Chicago Press, 1983) similarly points to Gadamer, Habermas, and Arendt as "possible avenues of inquiry" to an updated understanding of *phronesis*. Peter Steinberger presents Michael Oakeshott and Arendt as subscribing to Aristotle's concept of political judgment in *The Concept of Political Judgment* (Chicago: University of Chicago Press, 1993). And Joseph Dunne argues that the work of Gadamer, Habermas, and Arendt as well as John Henry Newman and R.G. Collingwood all can be related to Aristotle's concept of *phronesis* in *Back to the Rough Ground* (Notre Dame, IN: University of Notre Dame Press, 1993). More recently, Bent Flyvbjerg even suggests that the whole of the social sciences should become "*phronetic*" in *Making Social Science Matter* (Cambridge: Cambridge University Press, 2001). For a brief review of some other contemporary scholarship calling for a revival of *phronesis* see Richard S. Ruderman's "Aristotle and the Recovery of Political Judgment," *American Political Science Review* 91, no. 2 (June 1997): 409–20.

4 *After Virtue: A Study in Moral Theory* (South Bend, IN: University of Notre Dame Press, 2007), 259.

5 A definitive account of this debate can be found in Andrew Feenberg's *Questioning Technology* (New York: Routledge, 1999). He argues that we can think about technology in one of two ways: technology as autonomous, what he calls "essentialism," or technology as a tool, what he calls "constructivism." Feenberg himself promotes this second option and calls for new technologies that "respect the person," "create humane living spaces," and "mediate new social forms" ("From Essentialism to Constructivism," in *Technology and the Good Life?* ed. Eric Higgs, Andrew Light, and David Strong [Chicago: University of Chicago Press, 2000], 313). A similar idea is presented by Langdon Winner in *Autonomous Technology: Technics-out-of-control as a Theme in Political Thought* (Cambridge, MA: MIT Press, 1977); by Herbert Marcuse in *One-Dimensional Man* (Boston: Beacon Press, 1966); and by Emmanuel Mesthene in *Technological Change: Its Impact on Man and Society* (New York:

Signet, 1970). Essentialist approaches to technology include Marshall McLuhan's *Understanding Media: The Extensions of Man* (New York: McGraw-Hill, 1964), Jacques Ellul's *The Technological Society* (New York: Knopf, 1964), and most notably Martin Heidegger's essay "The Question Concerning Technology," in *Basic Writings*, ed. David Farrell Krell (San Francisco: HarperCollins, 1993), whose ideas will be discussed in more detail below.

1. Finding and Enforcing Limits

1 Mary Shelley, *Frankenstein or The Modern Prometheus* (London: George Routledge and Sons, 1888), 73.
2 Bill Joy, "Why the Future Doesn't Need Us: Our Most Powerful 21st Century Technologies – Robotics, Genetic Engineering, and Nanotech – Are Threatening to Make Humans an Endangered Species," *Wired*, April 2000: 238–62.
3 Bill Joy, "Act Now to Keep New Technologies Out of Destructive Hands," *New Perspectives Quarterly* 17, no. 3 (Summer 2000): 12–14.
4 "High Technology's Dark Side." 10 May 2000, available at http://www.pbs.org/newshour/bb/cyberspace/jan-june00/dark_5-10.html.
5 Available at http://www.carnegieendowment.org/publications/index.cfm?fa=view&id=17295.
6 *In the Matter of J. Robert Oppenheimer: The Security Clearance Hearing*, ed. Richard Polenberg (Ithaca: Cornell University Press, 2002), 46–7.
7 Hobbes writes: "But evil men, under pretext that God can do anything, are so bold as to say anything when it serves their turn, though they think it untrue; it is the part of a wise man to believe them no further than right reason makes that which they say appear credible. If this superstitious fear of spirits were taken away, and with it prognostics from dreams, false prophecies, and many other things depending thereon, by which crafty ambitious persons abuse the simple people, men would be much more fitted than they are for civil obedience" (*Leviathan* [London: Penguin, 1985], 93).
8 Ibid., 160.
9 Francis Bacon, *The New Organon and Related Writings*, ed. F.H. Anderson (New York: Macmillan, 1960), 23.
10 René Descartes, *Discourse on Method*, trans. Donald A. Cress (Indianapolis: Hackett, 1998), 35.
11 David Hume, *A Treatise of Human Nature*, vol. 1, ed. David Norton and Mary Norton (London: Oxford University Press, 2007), 266.
12 Marie Jean Antoine Nicolas de Caritat, marquis de Condorcet, *Sketch for a Historical Picture of the Progress of the Human Mind*, trans. June Barraclough (London: Weidenfeld & Nicolson, 1955).

13 August 2001.

14 Leon Kass, "Testimony in Front of the National Bioethics Advisory Commission," 14 March 1997, Watergate Hotel, Washington, DC. Available at bioethics.georgetown.edu/nbac/transcripts/1997/3-14-97.pdf.

15 Leon Kass, "Organs for Sale? Propriety, Property, and the Price of Progress," *The Public Interest*, no. 107 (Spring 1992): 86.

16 Leon Kass, "The New Biology: What Price Relieving Man's Estate?" *Science* 174, no. 401 (1971): 779–88. For more on Kass's considerable and often controversial work on bioethics see his *Toward a More Natural Science: Biology and Human Affairs* (New York: Free Press, 1985) and *Life, Liberty and the Defence of Dignity: The Challenge for Bioethics* (San Francisco: Encounter Books, 2004).

17 The passage is from an unpublished cycle of four lectures on technology Heidegger gave in 1949. It was first quoted in Wolfgang Schirmacher's *Technik und Gelassensheit* (Freiburg: Alber, 1983).

18 Martin Heidegger, "The Question Concerning Technology," in *Basic Writings*, ed. David Farrel Krell (San Francisco: HarperCollins, 1993), 332.

19 The Canadian philosopher George Grant suggests that ancient *techne* had a limited role, whereas technology is characterized by its complete lack of limitation (*Technology and Justice* [Concord, ON: Anansi, 1986], esp. 11–13). Along the same lines, Stanley Rosen argues that *techne* is defensive, whereas technology is offensive ("*Techne* and the Origins of Modernity," in *Technology in the Western Political Tradition*, ed. Arthur Melzer, Jerry Weinberger, and M. Richard Zinman [Ithaca: Cornell University Press, 1993], 73). Arthur Melzer proposes that rather than simply bringing something particular into being that would not have existed otherwise, as with *techne*, technology seeks to control nature as a whole ("The Problem with the 'Problem of Technology,'" ibid., 299). For these thinkers, contemporary technology is not simply more complicated or of a greater scope and size than ancient *techne*, but is fundamentally different.

20 Leon Kass, "The Problem of Technology," in *Technology in the Western Political Tradition*, ed. Melzer, Weinberger, and Zinman, 3–4. Kass repeats the same point in *Life, Liberty and the Defence of Dignity*: "Modern technology is less a bringing forth of objects than a setting upon, a challenging forth, a demanding of nature: that its concealed materials and energies be released and ordered as standing reserves, available and transformable for any multitude of purposes" ((32). For a discussion of the broader philosophical and religious ideas that help inform Kass's ideas see Lawrence Vogel's essay "Natural Law Judaism? The Genesis of Bioethics in Hans Jonas, Leo Strauss, and Leon Kass," in *Hastings Center Report* 36, issue 3 (May/June 2006): 32–44.

21 See Trish Glazebrook's excellent essay "From φύσισ to Nature, τέχνη to Technology: Heidegger on Aristotle, Galileo, and Newton," *Southern Journal of Philosophy* 38 (2000): 95–118, for a discussion of Heidegger's account of the relationship between nature and art.

22 Francis Fukuyama, *Our Posthuman Future: Consequences of the Biotechnology Revolution* (New York: Farrar, Straus, and Giroux, 2002), 182.

23 Ibid., 12.

24 Aristotle, *The Politics of Aristotle*, ed. and trans. Ernest Barker (London: Oxford University Press, 1958), book 7, chap. 14, sect. 8. Herein *Pol*. All citations of Aristotle's *Politics* will note the book, chapter, and section of this edition.

25 Aristotle, *The Nicomachean Ethics*, trans. W.D. Ross (London: Oxford University Press, 1998), 1141b20ff. Herein *NE*. All citations of Aristotle's *Ethics* will note the Bekker numbers.

26 *Pol* 7.1.13.

27 *Pol* 7.1.9.

28 For example, "providing a 'biologically related child' for an infertile or same sex couple; avoiding the risk of genetic disease; securing a genetically identical source of organs; 'replacing' a loved spouse or child who is dying or has died; or producing individuals of great genius, talent or beauty" (President's Council on Bioethics, *Human Cloning and Human Dignity: An Ethical Inquiry* [Washington, DC, 2002], 78.

29 For example, the council warns that reproductive cloning might lead to the breakdown of the family. They explain: "Procreation as traditionally understood invites acceptance, rather than reshaping, engineering, or designing the next generation. It invites us to accept limits to our control over the next generation" (ibid., 85).

30 Robert Wachbroit, "Genetic Encores: The Ethics of Human Cloning," *Report from the Institute of Philosophy and Public Policy* 17, no. 4 (1997), 2.

31 It has been argued that membership on the council was stacked against therapeutic cloning. See Chris Mooney, "Irrationalist in Chief," *The American Prospect* 12, no. 17 (24 September–8 October 2001); available at http://www.prospect.org/print/V12/17/mooney-c.html; and Stephen Hall, "Human Cloning: President's Bioethics Council Delivers," *Science* 297, issue 558 (2002): 322–4. Even when not explicit, as was the case with high-profile Bush supporters like Leon Kass, then the chairman of the council, and Charles Krauthammer, many members of the council had publicly spoken out against all forms of cloning. Council member Robert P. George, for example, wrote in the journal *National Review* that harvesting stem cells from human embryos is "grotesquely immoral" and decried any efforts

to publicly fund and promote this "injustice" (Robert P. George and Patrick Lee, "Reason, Science, & Stem Cells: Why Killing Embryonic Human Beings Is Wrong," *National Review Online*, 20 July 2001; available at http://www.nationalreview.com/comment/comment-george072001.shtml).

Council member Mary Ann Glendon is a signatory to the "Statement of the Catholic Leadership Conference on Human Cloning." It reads: "The CLC endorses the position of President George W. Bush which he stated in his first formal address to the American people: I strongly oppose human cloning, as do most Americans. We recoil at the idea of growing human beings for spare body parts or creating life for our convenience ... Even the most noble ends do not justify any means ... The moral justification of any research cannot be based upon the dehumanizing promise that a good end justifies the use of any means necessary. Destroying human life in order to help human life is intrinsically evil" (available at http://www.priestsforlife.org/articles/01-11-01humancloningclc.htm, November 2001).

Not long after the report's release, the members on the other side of the debate expressed frustration with the direction the council had taken. Janet D. Rowley, Elizabeth Blackburn, Michael S. Gazzaniga, and Daniel W. Foster, all traditional university scientists, objected to the moratorium. In an open letter, they wrote: "The President's Council, composed primarily of academics, now proposes to maintain our ignorance by preventing any research for four more years. That proposal is short-sighted: It will force U.S. scientists who have private funding to stop their research, and it will accelerate the brain drain to more enlightened countries ... Our ignorance in this vitally important area is profound, and the potential for meaningful medical advances is very high indeed. To realize that potential, we must remove the current impediments to this critical research. Scientists should become more active in urging Congress to lift the ban and to establish the proposed, broadly constituted regulatory board NOW" ("Harmful Moratorium on Stem Cell Research," *Science* 297 [2002]: 1957). The frustration of these scientists reflects a broader alienation of the scientific community from the administration. To add more fuel to the fire, on 27 February 2004, Professor Blackburn and William May were told their services would no longer be needed and were dismissed from serving on the council. Blackburn said she believed she was let go because her political views did not match those of the president and of Kass, with whom she had often been at odds at council meetings. "I think this is Bush stacking the council with the compliant," Blackburn said to the *Washington Post*. Three new members were named to take their places. They included a doctor who had called for more religion in public life, a political scientist who had spoken out

precisely against stem cell research, and another who had written about
the "threats of biotechnology" (Rick Weiss, "Bush Ejects Two from Bioethics Council: Changes Renew Criticism That the President Puts Politics
Ahead of Science," 28 February 2004: p. A6).

32 Many bioethicists seek an "ethical bypass" out of this debate. See, for example, Mary B. Mahowald and Anthony P. Mahowald, "Embryonic Stem
Cell Retrieval and a Possible Ethical Bypass," *American Journal of Bioethics*
2, no. 1 (2002): 42–3. A promising way out is adult stem cell research. Marilyn Coors writes: "The challenge lies in making cells derived from adult
stem cells function effectively. If this hurdle can be overcome, adult stem
cells promise to be a practical, efficient, and therapeutic option that avoids
the ethical problems associated with the therapeutic cloning" ("Therapeutic Cloning: From Consequences to Contradiction," *Journal of Medicine
and Philosophy* 27, no. 3 [2002]: 297–317). Harvesting adult stem cells from
blood, bone marrow, or tissue does not require the creation or destruction
of an embryo. Just as we give blood or tissue for medical tests for the benefit of our own health, we could provide stem cells for the development of
therapies and organs. But as of this writing, therapeutic cloning is the best
way to get stem cells.

A Brief Note on the Historical Review

1 Even though the great books of political philosophy may seem an unlikely
source for ideas about how to live better with technology, there is a rich
recent history of similar efforts to locate the source of our technological dilemma through an analysis of these texts.

For example, Jürgen Habermas has argued that Hobbes's *Leviathan*, written in the mid-seventeenth century, clearly indicates a major shift away
from the old Aristotelian model of politics based in "practical prudent action" to a new model based on the modern experimental sciences. Summarizing Hobbes's logic, he explains that: "With a knowledge of the general
conditions for a correct order of the state and of society, practical prudent
action of human beings toward each other is no longer required, but what is
required instead is the correctly calculated generation of rules, relationships
and institutions ... Human behaviour is therefore to be now considered
only as the material for science. The engineers of the correct order can dis-
regard the categories of ethical social intercourse and confine themselves to
the construction of conditions under which human beings, just like objects
within nature, will necessarily behave in a calculable manner. This separa-
tion of politics from morality replaces instruction in leading a good and just

life with making possible a life of well-being within a correctly instituted order" (*Theory and Practice*, trans. John Viertel [Boston: Beacon Press, 1973], 43).

Habermas goes on to lament the rise of the scientific control of modern society and the associated loss of the practical orientation of classical politics, wondering whether it can be redeemed in our day and age. Habermas's inclination, and the inclination of others who share this same approach, is to return to these texts to not only understand when and how we began to lose this practical capacity to judge, discuss, instruct, and learn what it is to live a good and just life, but also to comprehend the original articulation of this practice in the hope that we can remember and relearn it.

2. *Phronesis* vs. *Techne*

1 *NE* 1140b20.
2 *Pol* 1.11.8.
3 *Pol* 1.2.9.
4 *NE* 1140a10.
5 *NE* 1140b20.
6 Ronald Beiner, *Political Judgment* (Chicago: University of Chicago Press, 1983), 74.
7 Thucydides, *History of the Peloponnesian War*, trans. Rex Warner (London: Penguin, 1972), 117 (1.138).
8 It should be noted that Aristotle makes a distinction between *synesis* and *phronesis*, explaining that the former describes good advice whereas the latter describes good action (*NE* 1143a7–10). However, it is clear that Thucydides is using the word to describe Themistocles's "rapidity of action."
9 *NE* 1103a14ff.
10 *NE* 1142a12–21.
11 A similar rationale might be used to explain the minimum voting age requirement in today's democracies.
12 *NE* 1140a25.
13 *NE* 1106b36–7a2.
14 *NE* 1113a9–13.
15 *Pol* 7.14.15–17.
16 *Pol* 7.14.18.
17 *Pol* 7.14.8.
18 Ibid.
19 *Pol* 4.1.7–8.

20 261d.

21 "Kingly-*techne* blends and weaves together; taking on the one hand those whose nature tend rather to courage, which is the stronger element and may be regarded as the warp, and on the other hand those which incline to order and gentleness … the woof – these, which are naturally opposed, she seeks to bind and weave together" (line 309). We are told that the Statesman, by weaving the temperate with the courageous, can transform the vices into virtues and cure each side of its deficiencies (line 310).

22 In other words, for *phronesis*, the agent of change or the efficient cause is in the material and not an external source. Differently, in the case of *techne*, the agent of change or efficient cause is from an external source – in the *technites*, not the product.

23 *Physics* 192b, 14–15.

24 *NE* 1140a10–15.

25 *NE* 1140b6.

26 *Pol* 7.4.4.

27 *Pol* 7.12.9.

28 Furthermore, as he explains later in the *Politics*, we can only "pray that our state should be ideally equipped at all points where fortune is sovereign – as we assume her to be in the sphere of the 'given'" (7.13.9). The make-up of the population would likely be included in this sphere of the given rather than the made.

29 *Pol* 2.1.1.

30 *Pol* 7.13.13.

31 *Pol* 1.10.1–2.

32 *Pol* 3.15.7.

3. The Decline of Good Judgment

1 Aristotle also warns that too much change, too quickly can weaken the power of law: "To change the practice of an art is not the same as to change the operation of a law. It is from habit, and only from habit, that law derives the validity which secures obedience. But habit can be created only with the passage of time; and a readiness to change from existing to new and different laws will accordingly tend to weaken the general power of law" (*Pol* 2.9.19–24).

2 *Confessions*, book 11, chapter 26.

3 Letter 140.

4 "Of the Morals of the Catholic Church," 8.13.

5 Book 1, chapter 2.

6 Book 1, chapter 3.
7 Alasdair MacIntyre, *After Virtue: A Study in Moral Theory* (University of Notre Dame, 1984), 182.
8 See, for example, Aristotle, *NE* 4.1 and 2.
9 Making a similar point, in *Pagan and Christian in an Age of Anxiety* (Cambridge: Cambridge University Press, 1990) the classical scholar E.R. Dodds notes: "What astonished all the early pagan observers ... was the Christians' total reliance on unproved assertion – their willingness to die for the indemonstrable ... The Christians possess three of the four cardinal virtues: they exhibit courage, self-control and justice; what they lack is *phronesis*, intellectual insight, the rational basis of the other three" (121).
10 Consider 1 Corinthians 13:13: "There are three things that last forever: faith, hope, and love ... but the greatest of them all is love."
11 Anthony J. Celano, "The End of Practical Wisdom: Ethics as Science in the Thirteenth Century," *Journal of the History of Philosophy* 33, no. 2 (April 1995): 230.
12 *Summa theologiae* 2-2, 52.1.
13 Ibid., 2-2, 47.6.
14 Aquinas also writes, "The right ends of human life are fixed" (ibid., 47.15); that prudence has its source in an understanding of those ends (ibid., 49.2); and that "prudence, which denotes rectitude of reason, is chiefly perfected and helped through being ruled and moved by the Holy Ghost" (ibid., 52.2).
15 In third century AD Domitius Ulpianus, the Roman jurists explained: "*Jus naturale* is that which nature has taught to all animals; for it is not a law specific to mankind but to all animals – land animals, sea animals, and birds as well. Out of this comes the union of man and woman which we call marriage, and the procreation of children, and their rearing. So we can see that the other animals, wild beasts included, are rightly understood to be acquainted with this law. *Jus gentium*, the law of nations, is that which all human peoples observe. That it is not coextensive with natural law can be grasped easily, since this latter is common to all animals whereas *jus gentium* is common only to human beings among themselves." *The Digest of Justinian*, ed. T. Mommsen, A. Watson, and P. Krueger (Philadelphia: University of Pennsylvania Press, 1985), 1.1.1.3,4.
16 Aquinas writes: "The natural law is nothing else than the rational creature's participation of the eternal law" (*Summa theologiae* 2-1, 91.2).
17 Ibid., 2-2, 47.15.
18 Ibid., 2-2, 49.2.
19 Ibid., 2-2, 52.2.

20 Ibid., 2-1, 2.1–8.
21 Ibid., 2-1, 3.8.
22 Ibid., 2-1, 5.3.
23 Ibid., 2-1, 4.7 and 2-2, 184.2.
24 Ibid., 2-2, 55.1.
25 Ibid., 2-1, 4.5.
26 Ibid., 2-1, 4.8.
27 Book 1, 20.

4. The Rise of Technical Knowledge

1 A religious and political schism resulted in the fracturing of the Church's power in Europe. This period is often referred to as the Western Schism (1378–1417), which saw the election of three rival popes and thus three rival factions within the Church. The 1492 election of Pope Alexander VI, known for his corruption and numerous illegitimate children, further damaged the moral authority of the Church.
2 *The Prince*, chap. 25, p. 90. Quotes from Machiavelli are from *The Chief Works and Others: Volumes I and II*, trans. Allan H. Gilbert (Durham, NC: Duke University Press, 1989).
3 Machiavelli writes: "I compare Fortune with one of our destructive rivers which, when it is angry, turns the plains into lakes, throws down the trees and the buildings, takes earth from one spot, puts it in another; everyone flees before the flood; everyone yields to its fury and nowhere can repel it. Yet though such it is, we need not therefore conclude that when the weather is quiet, men cannot take precautions with both embankments and dykes, so that when the waters rise, either they go off by a canal or their fury is neither so wild nor so damaging. The same things happen about Fortune. She shows her power where strength and wisdom do not prepare to resist her, and directs her fury where she knows that no dykes or embankments are ready to hold her" (ibid., 25.90). Technical means, the building of defences and embankments, are applied to subdue nature or chance. And, while the river is redirected as it is moving through the city, its course is not fundamentally altered on the landscape as a whole. Machiavelli believes that a portion of nature can be controlled, but that it cannot be completely overcome.
4 Ibid., 3.16. Machiavelli often used the terms wisdom (*saggezza*) and prudence (*prudenza*) interchangeably.
5 Ibid., 7.62.
6 In *Discourse on Livy*, he explains: "Prudent men usually say (and not by chance or without merit) that whoever wants to see what is to be, considers

what has been; for all the things of the world in every time have had the very resemblance as those of ancient times. This arises because they are done by men who have been, and will always have, the same passions, and of necessity they must result in the same effects. It is true that men in their actions are more virtuous in this province than in another, according to the nature of the education by which those people have formed their way of living. It also facilitates the knowledge of future events from the past, to observe a nation hold their same customs for a long time, being either continuously avaricious, or continuously fraudulent, or have any other similar vice or virtù" (book 3, chap. 63).

7 *The Prince*, 15.58.
8 *Phys* 185a2ff.
9 *The Prince*, 6.25.
10 Thomas Hobbes, *The Leviathan*, ed. C.B. Macpherson (London: Penguin, 1985), 97.
11 Ibid.
12 Ibid., 117.
13 The Scientific Revolution demarcates a period during the sixteenth and seventeenth century that saw a great expansion in scientific knowledge, including proofs for heliocentricism, the birth of modern chemistry, and insights into the inner workings of the human body.
14 *The Leviathan*, 166.
15 Ibid., 308.
16 Ibid.
17 "For prudence is but experience, which equal time equally bestows on all men in those things they equally apply themselves unto" (ibid., 183). As he later says: "It is evident that we are not to account as any part thereof that original knowledge called experience, in which consisteth prudence, because it is not attained by reasoning, but found as well in brute beasts as in man; and is but a memory of successions of events in times past, wherein the omission of every little circumstance, altering the effect frustrateth the expectation of the most prudent: whereas nothing is produced by reasoning aright, but general, eternal, and immutable truth" (682).
18 Ibid., 392.
19 Ibid., 387.
20 Ibid., 223.
21 Ibid., 161–2.
22 Ibid., 379.
23 Ibid., 139.

24 Ibid., 263.
25 Ibid., 161–2.
26 Ibid., 264.
27 Ibid., 407–8.
28 This is similar to Machiavelli's "exhortation" to liberate Italy in chapter 26, the final chapter of *The Prince*. Both thinkers look to or, better yet, call to a great political leader to come forward and put their theories into practice.

5. After the Great Reversal

1 Abbot Lawrence, who founded the textile mills at Lawrence, Massachusetts, typified the common attitude among American industrialists when he remarked, "The water-power on the James River at Richmond is unrivalled," and lamented: "It seems a great waste of natural wealth to permit it to run into the sea, having hardly touched a water-wheel" (quoted in Theodore Steinberg, *Nature Incorporated* [Cambridge: Cambridge University Press, 2004], 71).
2 Stevens even hired a carnival performer known for swallowing and regurgitating stones to ingest silver spheres so he could later test the effect of the stomach's digestive juices upon them. See Thomas L. Hankin, *Science and the Enlightenment* (Cambridge: Cambridge University Press, 1985), 123.
3 As the historian Peter Gray points out, before the wide application of the Enlightenment belief in the power of reason to the human body, "It is safe to speculate that in the eighteenth century a sick man who did not consult a physician had a better chance of surviving than one who did" (*The Enlightenment: The Science of Freedom* [New York: Norton and Co., 1996], 19).
4 Joel Kupperman, *Character* (Oxford: Oxford University Press, 1991), 71–2.
5 *Groundwork of the Metaphysics of Morals*, in *Immanuel Kant: Groundwork of the Metaphysic of Morals in Focus*, ed. Lawrence Pasternack, trans. H.J. Patton (New York: Routledge, 2002), 39.
6 Immanuel Kant, *Foundations of the Metaphysics of Morals*, trans. Lewis White Beck (New York: Liberal Arts Press, 1959), 39 For a study of Kant's ideas on political judgment see Ronald Beiner, *Political Judgment* (Chicago: University of Chicago Press, 1983), 31–63. For a similar study that includes Marx see Dick Howard, *Political Judgments* (Lanham, MD: Rowman & Littlefield, 1996), 31–43 and esp. 133–51, 211–29.
7 Jean Bodin, *Six Books of The Commonwealth*, abr. and trans. M.J. Tooley (Oxford: Basil Blackwell, 1955), book 6, chap. 1.

8 In the same chapter, he decides that "If at any time they omitted the censorship, as occasionally happened during a long war, one can see at a
   glance how the morals of the people declined, and the commonwealth fell
   sick, like a body denied its customary purgations ..."
9 *Sketch for a Historical Picture of the Progress of the Human Mind*, Tenth epoch,
   "Future Progress of Mankind" (trans. June Barraclough [London: Weidenfeld & Nicolson, 1955]).
10 "Essay on the Application of Analysis to the Probability of Majority Decisions" (1785), in *Condorect: Selected Writings*, ed. K.M. Baker. Indianapolis:
   Bobbs-Merrill Co., 1976.
11 He even suggests that the institution of public education would alter the
   physical function of the brain and that this change would be passed from
   one generation to the next, "It is therefore simple enough to believe that
   if several generations of men receive an education directed toward a constant goal, if each of the individuals comprising them cultivates his mind
   by study, succeeding generations will be born with a greater propensity
   for acquiring knowledge and a greater aptitude to profit from it" (*On Public Instruction*, First memorandum, "The Nature and Purpose of Public Instruction," 1791, in *Condorect: Selected Writings*, ed. Baker).
12 Anne-Robert-Jacques Turgot, "Memorandum on Local Government"
   (1775), in *The Old Regime and the French Revolution*, ed. K.M. Baker (Chicago: University of Chicago Press, 1987), 99.
13 The Scottish historian Thomas Carlyle describes the hectic build-up to the
   "Feast of Reason" celebrating the Goddess in the third volume of his three-
   volume *The French Revolution: A History*, first printed in 1837:
      "Above all things, there come Patriotic Gifts, of Church-furniture.
   The remnant of bells, except for tocsin, descend from their belfries into
   the National melting-pot to make cannon. Censers and all sacred vessels are beaten broad; of silver, they are fit for the poverty-stricken Mint;
   of pewter, let them become bullets, to shoot the 'enemies du genre hu-
   main.' ... In all Towns and Townships as quick as the guillotine may
   go, so quick goes the axe and the wrench: sacristies, lutrins, altar-rails
   are pulled down; the Mass-Books torn into cartridge papers: men dance
   the Carmagnole all night about the bonfire ... This, accordingly, is what
   the streets of Paris saw: 'Most of these persons were still drunk, with
   the brandy they had swallowed out of chalices; eating mackerel on the
   patenas! Mounted on Asses, which were housed with Priests' cloaks ...
   Next came Mules high-laden with crosses, chandeliers, censers, holy
   water vessels ...' For the same day, while this brave Carmagnole-dance
   has hardly jigged itself out, there arrive Procureur Chaumette and

Municipals and Departmentals, and with them the strangest freight-age: a New Religion! Demoiselle Candeille, of the Opera, a woman fair to look upon, when well rouged; she borne on palanquin shoulder-high; with red woolen nightcap; in azure mantle; garlanded with oak; holding in her hand the Pike of Jupiter-Peuple, sails in: heralded by white young women girt in tricolor. Let the world consider it! This, O National Convention wonder of the universe, is our New Divinity; *Goddess of Reason, worthy, and alone worthy of revering"* (New York: American Book Exchange, 1881), 581–2.

14 See, for example, Norman Schofield, "The Intellectual Contribution of Condorcet to the Founding of the US Republic 1785–1800," *Social Choice and Welfare* 25, nos. 2–3 (2005): 303–18.

15 Carl von Linné (a.k.a. Linnaeus) wrote *A General System of Nature through the Three Grand Kingdoms of Animals, Vegetables, and Minerals: Classes, orders, genera, species, and varieties, with their habitation, manners, economy, structure and peculiarities* (1735), a periodic table of plants and animals that places human beings in a nondescript corner along with other primates.

16 *Popular Instructions on the Calculation of Probabilities*, trans. Richard Beamish (London: J. Weale, 1849), 107.

17 This quote is taken from the 1842 English translation titled *A Treatise on Man and the Development of His Faculties* (Edinburgh: William and Robert Chambers), repr. in "Quetelet on the Study of Man," *Population and Development Review* 22, no. 3 (Sept. 1996): 550.

18 Ibid., 551.

19 In an earlier work, he similarly concludes: "The power of man over animal life, in producing whatever varieties of form he pleases, is enormously great. It would seem as though the physical structure of future generations was almost as plastic as clay, under the control of the breeder's will. It is my desire to show ... that mental qualities are equally under control." Francis Galton, "Hereditary Talent and Character," *Macmillan's Magazine* no. 12 (June and August 1865): 157. As quoted in Ruth Schwartz Cowan, "Francis Galton's Statistical Ideas: The Influence of Eugenics," *Isis* 63, no. 4 (Dec. 1972): 510.

20 Ibid.

21 Schwartz Cowan points out: "If one assumes, as Galton did, that every organic character is produced by a number of genetic determinants, then the problem of reducing and combining meteorological data so as to be able to predict the weather in one time at one place is strikingly similar to the problem of predicting the characteristics of an offspring once the constitution of its parents is known" (ibid., 513).

6 Responses to the Great Reversal

1 For a thorough review of the Romantics' view of technology, see again Paul Cantor's "Romanticism and Technology: Satanic Verses and Satanic Mills," in *Technology in the Western Political Tradition*, ed. Arthur Melzer, Jerry Weinberger, and M. Richard Zinman (Ithaca: Cornell University Press, 1993).

2 Ernst Jünger, "Technology as the Mobilization of the World through the Gestalt of the Worker" (1932), repr. in Carl Mitcham and Robert Mackey, eds, *Philosophy and Technology: Readings in the Philosophical Problems of Technology* (New York: Free Press, 1983).

3 Oswald Spengler, *Der Mensch und die Technik: Beitrag zu einer Philosophic des Lebens* (1931) (Munich: C.H. Beck'sche Verlagsbuchhandlung, 1971), as quoted in John Farrenkopf, "Spengler's Historical Pessimism and the Tragedy of Our Age," *Theory and Society* 22, no. 3 (1993): 405–6.

4 Martin Heidegger, "The End of Philosophy and the Task of Thinking," in *Basic Writings*, ed. David Farrel Krell (San Francisco: HarperCollins, 1993), 434.

5 Robert Bernasconi has already discussed a connection between authenticity and *phronesis*. He questions the legitimacy of a parallel between the dichotomies of *phronesis* and techne and authenticity and inauthenticity. R. Bernasconi, "Heidegger's Destruction of *Phronesis*," *Southern Journal of Philosophy* 28, supplement (1989): 127–47.

6 Martin Heidegger, *Plato's Sophist*, trans. Richard Rojecwicz and André Schuwer (Bloomington, IN: Indiana University Press, 1997), §8a (p. 47–9).

7 The familiar English word "authentic," associated with things like integrity and genuineness, comes from the Greek "authentikos," meaning original or authoritative. However, Heidegger does not use the German word "authentische," but instead the coined word "eigentlichkeit," which translates as something close to "ownmostness" or "that which is my own" (*eigen*), Heidegger presents authenticity in opposition to *Uneigentlichkeit*, inauthenticity, or "that which is not my own" (*uneigen*). See *Being and Time*, division 1, section 9.

8 *Plato's Sophist*, §8b (51–2).

9 Ibid.

10 Aristotle recognizes the problem of a corrupt education system perpetuating a corrupt political system. He points to the "vulgar decline" of statesmen who are concerned only with the "useful" and "profitable" as well as empire building (*Pol* 7.14.15–17). If the *polis* fails to pass down the bases of its *ethos*, laws, and the good life to the next generation of

citizens, then the constitution of the whole city will become deviant. He is clear that the citizens reproduced by that city's legislators will be unhappy (7.14.18).

11 Heidegger writes: "A future thinker, who is perhaps given the task of taking over this thinking which I have tried to prepare, will have to acknowledge the following words, which Heinrich von Kleist once wrote: 'I step back before one, who is not yet here, and I bow a millennium ahead of him, before his spirit.'" M. Heidegger, *Martin Heidegger in Conversation*, ed. Richard Wisser (Freiburg: Arnold-Heinemann, 1970), 47.

12 See "The Self-Assertion of the German University," in *The Heidegger Controversy: A Critical Reader* ed. Richard Wolin, trans. William S. Lewis (Cambridge: MIT Press, 1993), esp. 33.

13 Martin Heidegger, *An Introduction to Metaphysics*, trans. Ralph Manheim (New Haven: Yale University Press, 1959), 37–8.

14 Michael Gillespie explains that "Heidegger was attracted to Nazism because he believed it offered a solution to the crisis of Western civilization … Heidegger clearly felt that resolute action was needed to deal with the social and spiritual crisis and was attracted to the Nazis because of their determination for action." M. Gillespie, "Martin Heidegger's Aristotelian National Socialism," *Political Theory* 28, no. 2 (2000): 141–2.

15 See Martin Heidegger, *Being and Time*, trans J. Macquarrie and E. Robinson (San Francisco: Harper & Row, 1962), §§ 54–60.

16 Heidegger, "The Self-Assertion of the German University," 33.

17 Rather than the instrumental rationality, sterility, and humanism indicative of Plato's theory of the forms (eidos), Heidegger wanted to somehow recover a lost Hericlitean universe in flux where man is tossed "back and forth between structure and the structureless, order and mischief, between the evil and noble" (*Introduction to Metaphysics*, trans. Richard Polt and Gregory Fried [New Haven: Yale University Press, 2000], 161). He also asks, "But if that which is an essential consequence is raised to the level of essence itself, and thus takes the place of the essence, then how do things stand?" He continues, "What remains decisive is not the fact in itself that *phusis* was characterized as idea, but that the idea rises up as the solid and definitive interpretation of Being" (194). Heidegger explains that the idea or eidos is initially understood as the visible appearance of the "movedness" or "emerging power" of nature (*physis*). From here, *physis* as movedness is ignored in lieu of the superficial, unmoving eidos. Eidos becomes a paradeigma, a model or prototype rather than anything immediately apparent. Heidegger concludes, "Because the actual repository of being is the idea and this is the prototype, all

disclosure of being must aim at assimilation to the model, accommodation to idea" (184–5).

18 Steven Crowell argues that resolve can "be encompassed by no rules, assessed by no public criteria, be integrated into no public practices; it is not a form of skillful coping and cannot be thought of in terms of *phronesis*." In fact, we can hardly even say that resolve involves action because it "transpires on the basis of death, the total breakdown of such abilities-to-be." "Authentic Historicality," in *Space, Time, and Culture*, ed. David Carr and Canhui Zhang (Dordrecht, The Netherlands: Kluwer, 2004), 68.

19 *NE* 1094a22–6.

20 Wolin notes that Eric Weil and Alphonse de Waelhens support this view. On the one hand, Heidegger's many speeches and activities during his time as rector of Freiberg University made it clear that he promoted support of Hitler and National Socialism as "an affirmation of 'authentic existence.'" On the other hand, these same speeches and actions seemed not to explicitly support the Jewish and racial considerations normally associated with Nazism (Karl Löwith, "My Last Meeting with Heidegger in Rome, 1936," repr. in *The Heidegger Controversy: A Critical Reader*, ed. Wolin, 180–2). It seems possible that Heidegger truly believed that the Germany of the early 1930s held out the possibility for authentic existence – he was not simply swept up in the fervour of the times – but his vision of a Nazi state did not match Hitler's vision. Jeffrey Herf argues that the Nazi appropriation of the language of authenticity highlights the disjuncture between Heidegger's philosophy and his "political error" (Jeffrey Herf, *Reactionary Modernism: Technology, Culture and Politics in Weimar and the Third Reich* [Cambridge: Cambridge University Press, 1984], 224). Still, an argument can be made that Heidegger both explicitly (Tom Rockmore, *The Heidegger Case: On Philosophy and Politics* [Philadelphia: Temple University Press, 1992], 111) and implicitly supported Nazi racial considerations (ibid., 192). What Steiner calls Heidegger's "total public silence" (George Steiner, *Martin Heidegger* [Chicago: University of Chicago Press, 1987], 116) on the Holocaust and the policies of the Third Reich lends credence to this view. But, in a return letter to Herbert Marcuse written not long after the war, Heidegger addresses his failure to "provide a public, readily comprehensible counter-declaration." As he explains, "It would have been the end of both me and my family" (20 January 1948, repr. in *The Heidegger Controversy*, ed. Wolin, 163). This seems a most common and legitimate excuse for silence. It also worth reviewing some of the other points raised in the same letter: "I expected from National Socialism a spiritual renewal of life in its entirety"; "I recognized my political error and resigned my rectorship in protest against state and party"; "[I] was exploited for propaganda purposes

both here and abroad"; none of his students "fell victim to Nazi ideology"; "the bloody terror of the Nazis in point of fact had been kept secret from the German people." We can conclude with some confidence that Heidegger's philosophy did not match up with the politics of the day. This is evident in his Bremen lectures of 1949, where he compares the Holocaust to mechanized agriculture and nuclear war. It seems clear that Heidegger understands the Holocaust as one of the worst reflections of global technology – far from the "spiritual renewal" he was looking for.

21 Nicols Fox explains that "[Neo-]Luddism is neither conservative nor liberal: both capitalism and Marxism are committed to the concept of industrial progress, the wisdom of which Luddites question" (*Against the Machine: The Hidden Luddite Tradition in Literature, Art, and Individual Lives* [Washington, DC: Island Press, 2002], xvii).

22 This is Alston Chase's contention in his June 2000 *Atlantic Monthly* article "Harvard and the Making of the Unabomber," vol. 285, no. 6: 41–65.

23 Edward Abbey, *The Monkey Wrench Gang* (Philadelphia: Lippincott, 1975), 100–1.

24 Chuck Palahniuk, *Fight Club* (New York: Norton), 124.

25 Kirkpatrick Sale, "Lessons from the Luddites: Setting Limits on Technology," *The Nation* 260, no. 22 (1995): 785.

26 Chellis Glendinning, "Notes toward a Neo-Luddite Manifesto," *Utne Reader* 38, no. 1 (1990): 50.

27 Martin Heidegger, *Discourse on Thinking*, trans. John M. Anderson and E. Hans Freund (New York: Harper & Row, 1966), 19.

28 Martin Heidegger, *Martin Heidegger in Conversation*, ed. Richard Wisser (New Delhi: Arnold-Heinemann, 1970), 43. In his 1949 essay "The Turning," Heidegger unequivocally states that he is not advocating anything as ridiculous as the abandonment of technology. "Technology," [Heidegger] repeats, "will not be done away with. Technology will not be struck down, and certainly it will not be destroyed." As quoted in Iain Thomson, *Heidegger on Ontotheology* (Cambridge: Cambridge University Press, 2005), 72–3.

29 Olafson explains that "Heidegger appears to have understood Nazism as a way of having things both ways." That is to say, he embraced the idea that the Nazis would wipe away the inauthenticity of modern society while, at the same time, protect the authentic *völkisch* traditions of Germany by the military and economic power of a modern state. Frederick A. Olafson, "Heidegger's Thought and Nazism," *Inquiry* 43, no. 3 (2000): 277–8.

30 Heidegger, *The Heidegger Controversy*, ed. Wolin, 31.

31 Martin Heidegger, *Introduction to Metaphysics*, trans. Richard Polt and Gregory Fried (New Haven: Yale University Press, 2000), 174.

32 Thomson discusses the mislabelling of Heidegger as nostalgic for the pre-industrial world, as embracing the "reactionary antimodernism of a philosophical 'redneck'" or a "Luddite 'technophobia.'" Thomson, *Heidegger on Ontotheology*, 45.

33 Heidegger, *The Heidegger Controversy*, ed. Wolin, 32.

34 Heidegger, *Basic Writings*, ed. Krell, 362.

35 Ibid., 355.

36 Hubert Dreyfus and Charles Spinosa, "Highway Bridges and Feasts: Heidegger and Borgmann on How to Affirm Technology," *Man and World* 30 (1997): 173. See also Thomson, *Heidegger on Ontotheology*, 439. The "focal" is in reference to Albert Borgmann's work on focal things and practices (discussed below).

37 Dreyfus and Spinosa, "Highway Bridges and Feasts," 159.

38 Heidegger, *Discourse on Thinking*.

39 Later thinkers such as Neil Postman, Hans Jonas, Langdon Winner, and Albert Borgmann attempt a similar moderate response to the challenge of technology. For example, in *Technology and the Character of Contemporary Life*, Albert Borgmann writes: "Focal things and practices can empower us to propose and perhaps to enact a reform of technology" (*Technology and the Character of Contemporary Life* [Chicago: University of Chicago Press, 1984], 155). He then goes on to suggest that things such as cyclotrons and space shuttles may bear resemblance to medieval cathedrals and monuments in that they serve as "focal points" for our communities, inspiring awe and appreciation for the place of humanity in the cosmos: we can find peace and serenity in "midst of our own creations which surround us daily" (161). Borgmann and other moderate essentialists believe that by reorienting or reforming the way we relate to technology, by recognizing that it is revealing something to us, we can mitigate its threat. So yet again we see essentialism as something other than determinism. By recognizing and changing our relationship to technology, we can help determine the course technology will take.

40 *The Heidegger Controversy*, ed. Wolin, 111. The rest of the quote reads: "It seems to me that you are taking technology too absolutely. I do not see the situation of man in the world of global technology as a fate which cannot be escaped or unraveled. On the contrary, I see the task of thought to consist in helping man in general, within the limits allotted to thought, to achieve an adequate relationship to the essence of technology. National Socialism, to be sure, moved in this direction. But those people were far too limited in their thinking to acquire an explicit relationship to what is really happening today and has been underway for three centuries." In his book

On Heidegger's Nazism and Philosophy (Berkeley: University of California Press, 1997), Tom Rockmore explains: "Here, in his own way, Heidegger is signaling, as clearly as he can – candidly, and accurately – that his theory of technology is meant to carry out the ideas which the National Socialists were too limited to develop through a theory of technology with political consequences" (206).

41 As quoted in *The Heidegger Controversy*, ed. Wolin, 104.

42 As Zimmerman puts it, "Despite his descriptions of how the old world was being obliterated by the advance of the technological one, Heidegger did not finally despair. Rather, he held out the hope that a saving power could grow from out of the dangerous depths of technological nihilism." Michael Zimmerman, *Heidegger's Confrontation with Modernity* (Bloomington: Indiana University Press, 1990), 133. Zuckert similarly explains, "What he had learned both from his study of the history of philosophy and the outcome of World War II was the impossibility of checking this technological leveling with 'will' or force." Catherine Zuckert, "Martin Heidegger: His Philosophy and His Politics," *Political Theory* 18, no. 1 (1990): 72.

43 *Basic Writings*, ed. Krell, 341.

44 Lewis Mumford, *Pentagon of Power: The Myth of the Machine*, vol. 2 (New York: Harcourt Brace Jovanovich, 1974), 433.

45 Jacques Ellul, *The Technological Society* (New York: Vintage, 1967), xxxiii.

46 Ibid., 311.

47 Ibid., 316.

48 Marshall McLuhan, *The Gutenberg Galaxy* (Toronto: University of Toronto Press, 1992), 3. McLuhan is not calling for the elimination or destruction of technology. In a late interview, he puts it bluntly, "Resenting a new technology will not halt its progress." M. McLuhan, "Playboy Interview," in *Essential McLuhan*, ed. Eric McLuhan and Frank Zingrone (Concord, ON: Anansi, 1995). He continues: "First of all – and I'm sorry to repeat this disclaimer – I'm not advocating anything; I'm merely probing and predicting trends. Even if I opposed them or thought them disastrous, I couldn't stop them, so why waste my time lamenting? ... I see no possibility of a worldwide Luddite rebellion that will smash all machinery to bits, so we might as well sit back and see what is happening and what will happen to us ... The central purpose of all my work is to convey this message, that by understanding media as they extend man, we gain a measure of control over them ... If we persist, however, in our rearview-mirror approach to these cataclysmic developments, all of Western culture will be destroyed and swept into the dustbin of history" (264–5).

49 George Grant, *Technology and Empire* (Toronto: Anansi, 1969), 139.
50 Ibid.
51 Heidegger, "The Question Concerning Technology," in *Basic Writings*, ed. Krell, 340.

7. Virtue in a Technological Age

1 An 1862 issue of the leading medical journal *The Lancet* speaks of "the impression on the brain produced by the rattle affecting it through the nerves of hearing; the rapid succession of objects presented to the sight, and the vibrations actually transmitted by the movement of the carriage to the very substance of the brain and spinal cord ... the effects of hurry and anxiety to catch trains, and the frequent concentration of effort required to compress business matters at the last moment into the strict limits imposed." John Ruskin, the nineteenth-century social critic and poet, warned of a different threat posed by trains, forewarning of a "deterioration of moral character in the inhabitants of every district penetrated by the railway." John Ruskin, "A Protest against the Extension of Railways in the Lake District" (1876), in *The Works of John Ruskin*, ed. E.T. Cook and Alexander Wedderburn, 39 vols. (London: George Allen, 1903–12), 34: 141, as quoted in Ralph Harrington, "The Neuroses of the Railway," *History Today* 44, no. 7 (1994).
2 In his essay "What Is Practice? The Conditions of Social Reason," Gadamer writes that "in our civilization, characterized by technological growth, what has been artificially produced sets the new terms" and, in turn, there is "the loss of flexibility in our interchange with the world" (71). He argues that "it is inevitable, then, that the modern technology of communication leads to a more powerful manipulation of our minds" and "the individual in society who feels dependent and helpless in the face of its technically mediated life forms becomes incapable of establishing an identity" (73) (in *Reason in the Age of Science*, trans. F.G. Lawrence [Cambridge: MIT Press, 1981]).
3 *Pol* 7.1.9.
4 Peter D. Kramer, *Listening to Prozac* (New York: Penguin Books, 1993), 13.
5 Ibid., 32.
6 S.B. Patten, J.V. Williams, J. Wang, C.E. Adair, R. Brant, A. Casebeer, and C. Barbui, "Antidepressant Pharmacoepidemiology in a General Population Sample," *Journal of Clinical Psychopharmacology* 25, no. 3 (2005): 285–7. See also National Center for Health Statistics, United States, *Chartbook on Trends in the Health of Americans* (Hyattsville, MD, 2007), 88.
7 This problem of circumvention applies to many new drugs. The development of drugs like Viagra suggests that such things as sexual arousal can be regulated by external means.

8 In a 1999 report, the International Narcotics Control Board (INCB), an agency of the World Health Organization, noted: "In the Americas, particularly in the United States, performance enhancing drugs are given to children to boost school performance or help them conform with the demands of school life."

9 See Michael S. Bahrke, Charles E. Yesalis III, and James E. Wright, "Psychological and Behavioural Effects of Endogenous Testosterone Levels and Anabolic-Androgenic Steroids among Males: A Review," *Sports Medicine* 10, no. 5 (1990): 303–37.

10 Amy Shipley, "Chemists Stay a Step Ahead of Drug Testers," *Washington Post*, 18 October 2005, p. E01.

11 Steve Gardner, "Young Players Need a Few Good Role Models," *USA Today*, 28 March 2005. President Barack Obama expressed a similar sentiment in his 9 February 2009 press conference: "What I'm pleased about is Major League Baseball seems to finally be taking this seriously, to recognize how big a problem this is for the sport, and that our kids hopefully are watching and saying, 'You know what? There are no short cuts, that when you try to take short cuts, you may end up tarnishing your entire career, and that your integrity's not worth it.'"

12 M. Thevis and W. Schänzer, "Emerging Drugs – Potential for Misuse in Sport and Doping Control Detection Strategies," *Mini-Reviews in Medicinal Chemistry* 7, no. 5 (2007): 531–7.

13 At http://www.plenitas.com/en/plastic-surgery-video-testimonials.htm.

14 See Edward H. Livingston, "Development of Bariatric Surgery-Specific Risk Assessment Tool," *Surgery for Obesity and Related Diseases* 3, no. 1 (January 2007): 14–20.

15 See A. Towbin, T. Inge, V. Garcia, H. Roehrig, R. Clements, C. Harmon, and S. Daniels, "Beriberi after Gastric Bypass Surgery in Adolescence," *Journal of Pediatrics* 145, no. 2 (2004): 263–7.

16 George Dvorsky, "Better Living through Transhumanism," *Humanist* 64, no. 3 (May 2004): 7–10.

17 Julian Huxley, *Religion without Revelation* (London: E. Benn, 1927).

18 Julian Huxley, "Transhumanism" (1957), repr. in *Journal of Humanistic Psychology* 8 (1968): 74–5.

19 Francis Fukuyama, "Transhumanism," *Foreign Policy*, no. 144 (Sept./Oct. 2004): 42–3.

20 *Leviathan*, 407–8.

Bibliography

Abbey, Edward. 1975. *The Monkey Wrench Gang*. Philadelphia: Lippincott.

Alper, Matthew. 1996. *The "God" Part of the Brain: A Scientific Interpretation of Human Spirituality and God*. Naperville, IL: Sourcebooks, Inc.

Aquinas, St Thomas. 1947. *Summa theologiae*, trans. English Dominican Fathers. New York: Benziger Brothers.

Aquinas, St Thomas. 2007. *Commentary on Aristotle's Politics*. Indianapolis: Hackett Publishing Inc.

Arendt, Hanna. 1998. *The Human Condition*. 2nd ed. Chicago: University of Chicago Press.

Aristotle. 1930. *Physica*, trans. R.P. Hardie and R.K. Gaye. In *The Works of Aristotle*, vol. 2. Oxford: Clarendon Press.

Aristotle. 1958. *The Politics of Aristotle*, ed. and trans. Ernest Barker. London: Oxford University Press.

Aristotle. 1998. *The Nicomachean Ethics*, trans. W.D. Ross. London: Oxford University Press.

Aristotle. 2002. *Metaphysics*. Santa Fe, NM: Green Lion Press.

Augustine, St. 1994. "Of the Morals of the Catholic Church," trans. Richard Stohert. In *The Nicene and Post-Nicene Fathers*, ed. Philip Schaff, new series, vol. 4. Peabody, MA: Hendrickson Publishers.

Augustine, St. 2003. *Concerning the City of God against the Pagans*. Trans. Henry Bettenson. London: Penguin Books.

Augustine, St. 2008. *The Confessions*, trans. Henry Chadwick. Oxford: Oxford University Press.

Bacon, Francis. 1960. *The New Organon and Related Writings*, ed. F.H. Anderson. New York: Macmillan.

Bahrke, Michael S., Charles E. Yesalis III, and James E. Wright. Nov 1990. "Psychological and Behavioural Effects of Endogenous Testosterone Levels and Anabolic-Androgenic Ateroids among Males. A Review." *Sports Medi-cine* (Auckland, NZ) 10, no. 5: 303–37. http://dx.doi.org/10.2165/00007256-199010050-00003. Medline:2263798.

Beiner, Ronald. 1983. *Political Judgment*. Chicago: University of Chicago Press.

Berlin, Isaiah. 1969. "Two Concepts of Liberty." In *Four Essays on Liberty*. Oxford: Oxford University Press.

Bernasconi, Robert. 1989. "Heidegger's Destruction of *Phronesis*." *Southern Journal of Philosophy* 28 (Supplement): 127–47.

Bernstein, Richard. 1983. *Beyond Objectivism and Relativism: Science, Hermeneutics, and Praxis*. Philadelphia: University of Pennsylvania Press.

Blake, William. 1982. "Preface: Milton." In *The Complete Poetry and Prose of William Blake*, ed. David V. Erdman, 95–6. Berkeley: University of California Press.

Bodin, Jean. 1955. *Six Books of the Commonwealth*, trans. M.J. Tooley. Oxford: Basil Blackwell.

Borgmann, Albert. 1984. *Technology and the Character of Contemporary Life*. Chicago: University of Chicago Press.

Cantor, Paul. 1993. "Romanticism and Technology: Satanic Verses and Satanic Mills." In *Technology in the Western Political Tradition*, ed. Arthur Melzer, Jerry Weinberger, and M. Richard Zinman, 109–28. Ithaca: Cornell University Press.

Carlyle, Thomas. 1881. *The French Revolution: A History*. New York: American Book Exchange.

Celano, Anthony J. 1995. "The End of Practical Wisdom: Ethics as Science in the Thirteenth Century." *Journal of the History of Philosophy* 33, no. 2: 225–43.

Chase, Alston. 2000. "Harvard and the Making of the Unabomber." *Atlantic Monthly* 285, no. 6: 41–65.

Condorcet, J.-A.-N. C. 1955. *Sketch for a Historical Picture of the Progress of the Human Mind*, trans. June Barraclough. London: Weidenfeld & Nicolson.

Condorcet, J.-A.-N. C. 1976. "Essay on the Application of Mathematics to the Theory of Decision-Making." In *Condorcet: Selected Writings*, ed. K.M. Baker, 33–70. Indianapolis: Bobbs-Merrill Co.

Condorcet, J.-A.-N. C. 1976. "The Nature and Purpose of Public Instruction." In *Condorect: Selected Writings*, ed. K.M. Baker, 105–42. Indianapolis: Bobbs-Merrill Co.

Condorcet, J.-A.-N. C. 1979. *Sketch for a Historical Picture of the Progress of the Human Mind*. Westport, CT: Greenwood Press.

Coors, Marilyn. 2002. "Therapeutic Cloning: From Consequences to Contradiction." *Journal of Medicine and Philosophy* 27, no. 3: 297–317. http://dx.doi.org/10.1076/jmep.27.3.297.2985. Medline:12187436.

Crowell, Steven. "Authentic Historicality." In *Space, Time, and Culture*, ed. David Carr and Canhui Zhang, 57–72. Dordrecht: Kluwer.

Descartes, René. 1998. *Discourse on Method*, trans. Donald A. Cress. Indianapolis: Hackett.

de Waelhens, Alphonse. 1947. "La philosophie de Heidegger et le nazisme." *Les Temps Modernes* 3: 115–27.

Dodds, E.R. 1990. *Pagan and Christian in an Age of Anxiety*. Cambridge: Cambridge University Press.

Dreyfus, Hubert, and Charles Spinosa. 1997. "Highway Bridges and Feasts: Heidegger and Borgmann on How to Affirm Technology." *Man and World* 30, no. 2: 159–77. http://dx.doi.org/10.1023/A:1004299524653.

Dunne, Joseph. 1993. *Back to the Rough Ground*. Notre Dame: University of Notre Dame Press.

Dvorsky, George. 2004. "Better Living through Transhumanism." *Humanist* 64, no. 3 (May): 7–10.

Einstein, Albert. 1939. "Einstein's Nuclear Warning." Letter written to President Roosevelt. http://www.carnegieendowment.org/2005/08/03/einstein-s-nuclear-warning/7hf.

Ellul, Jacques. 1964. *The Technological Society*. New York: Knopf.

Ellul, Jacques. 1967. *The Technological Society*. New York: Vintage.

Ellul, Jacques. 1980. *The Technological System*. New York: Continuum Publishing.

Farrenkopf, John. 1993. "Spengler's Historical Pessimism and the Tragedy of Our Age." *Theory and Society* 22, no. 3: 391–412. http://dx.doi.org/10.1007/BF00993534.

Feenberg, Andrew. 1999. *Questioning Technology*. New York: Routledge.

Feenberg, Andrew. 2000. "From Essentialism to Constructivism: Philosophy of Technology at the Crossroads." In *Technology and the Good Life?* ed. Eric Higgs, Andrew Light, and David Strong, 294–315. Chicago: University of Chicago Press.

Flyvbjerg, Bent. 2001. *Making Social Science Matter*. Cambridge: Cambridge University Press.

Foot, Phillipa. 1978. "Virtues and Vices." In *Virtues and Vices and Other Essays in Moral Philosophy*, 1–18. Berkeley: University of California Press.

Fox, Nicols. 2002. *Against the Machine: The Hidden Luddite Tradition in Literature, Art, and Individual Lives*. Washington, DC: Island Press.

Fukuyama, Francis. 2002. *Our Posthuman Future: Consequences of the Biotechnology Revolution*. New York: Farrar, Straus, and Giroux.

Fukuyama, Francis. 2004. "Transhumanism." *Foreign Policy* 144 (September/October): 42–3. http://dx.doi.org/10.2307/4152980.

Gadamer, Hans-Georg. 1981. "What Is Practice? The Conditions of Social Reason." In *Reason in the Age of Science*, trans. Frederick G. Lawrence, 69–87. Cambridge: MIT Press.

Galton, Francis. 1865. "Hereditary Talent and Character." *Macmillan's Magazine*, June and August, 12.

Galton, Francis. 1869. Repr. 2010. *Hereditary Genius*. Memphis, TN: General Books, Macmillan.

Gardner, Steve. 2005. "Young Players Need a Few Good Role Models." *USA Today*, 29 March. http://www.usatoday.com/sports/baseball/columnist/gardner/2005-03-28-gardner_x.htm.

George, Robert P., and Patrick Lee. 2001. "Reason, Science, & Stem Cells: Why Killing Embryonic Human Beings Is Wrong." *National Review*, 20 July.

Gillespie, Michael. 2000. "Martin Heidegger's Aristotelian National Socialism." *Political Theory* 28, no. 2: 140–66. http://dx.doi.org/10.1177/0090591700028002002.

Glazebrook, Trish. 2000. "From φύσισ to Nature, τέχνη to Technology: Heidegger on Aristotle, Galileo, and Newton." *Southern Journal of Philosophy* 38, no. 1: 95–118. http://dx.doi.org/10.1111/j.2041-6962.2000.tb00892.x.

Glendinning, Chellis. 1990. "Notes toward a Neo-Luddite Manifesto." *Utne Reader* 38, no. 1: 50–3.

Grant, George. 1969. *Technology and Empire*. Toronto: Anansi.

Grant, George. 1986. *Technology and Justice*. Concord, ON: Anansi.

Gray, Peter. 1996. *The Enlightenment: The Science of Freedom*. New York: Norton and Co.

Habermas, Jürgen. 1973. *Theory and Practice*, trans. John Viertel. Boston: Beacon Press.

Hall, Stephen. 2002. "Human Cloning: President's Bioethics Council Delivers." *Science* 297, no. 5580: 322–4. Medline:12130762

Hankin, Thomas L. 1985. *Science and the Enlightenment*. Cambridge: Cambridge University Press.

Harrington, Ralph. 1994. "The Neuroses of the Railway." *History Today* 44, no. 7: 15–21.

Heidegger, Martin. 1959. *An Introduction to Metaphysics*, trans. Ralph Manheim. New Haven: Yale University Press.

Heidegger, Martin. 1962. *Being and Time*, trans. J. Macquarrie and E. Robinson. San Francisco: Harper & Row.

Heidegger, Martin. 1966. *Discourse on Thinking*, trans. John M. Anderson and E. Hans Freund. New York: Harper & Row.

Heidegger, Martin. 1970. *Martin Heidegger in Conversation*, ed. Richard Wisser. Freiburg: Arnold-Heinemann.

Heidegger, Martin. 1993. "Building Dwelling Thinking." In *Basic Writings*, ed. David Farrel Krell. San Francisco: HarperCollins.

Heidegger, Martin. 1993. "The End of Philosophy and the Task of Thinking." In *Basic Writings*, ed. David Farrel Krell. San Francisco: HarperCollins.

Heidegger, Martin. 1993. "The Question Concerning Technology." In *Basic Writings*, ed. David Farrell Krell. San Francisco: HarperCollins.

Heidegger, Martin. 1993. "The Self-Assertion of the German University" [the rectoral address]. In *The Heidegger Controversy: A Critical Reader*, ed. Richard Wolin, trans. William S. Lewis. Cambridge, MA: MIT Press.

Heidegger, Martin. 1997. *Plato's Sophist*, trans. Richard Rojecwicz and André Schuwer. Bloomington: Indiana University Press.

Heidegger, Martin. 2000. *Introduction to Metaphysics*, trans. Richard Polt and Gregory Fried. New Haven: Yale University Press.

Herf, Jeffrey. 1984. *Reactionary Modernism: Technology, Culture and Politics in Weimar and the Third Reich*. Cambridge: Cambridge University Press.

Hobbes, Thomas. 1985. *The Leviathan*, ed. C.B. Macpherson. London: Penguin.

Howard, Dick. 1996. *Political Judgments*. Lanham, MD: Rowman & Littlefield.

Hume, David. 2007. *A Treatise of Human Nature*, vol. 1, ed. David Norton and Mary Norton. London: Oxford University Press. http://dx.doi.org/10.1522/25022634.

Hursthouse, Rosalind. 1999. *On Virtue Ethics*. Oxford: Oxford University Press.

Huxley, Aldous. 1998. *Brave New World*. New York: Perennial Classics, Harper Collins Publishers.

Huxley, Aldous. 2004. *The Doors of Perception and Heaven and Hell*. New York: Perennial Classics, HarperCollins Publishers.

Huxley, Julian. 1927. *Religion without Revelation*. London: E. Benn.

Huxley, Julian. 1957. "Transhumanism." Reprinted in *Journal of Humanistic Psychology* 8 (1968): 74–5.

"The Influence of Railway Travelling on Public Health." 1862. *The Lancet* 1: 79–84.

International Narcotics Control Board (INCB). 1999. "Europeans Taking 'Downers.' Americans Taking 'Uppers.'" INCB Annual Report 1998, release no. 4. Vienna, Austria.

Joy, Bill. 2000. "Act Now to Keep New Technologies Out of Destructive Hands." *New Perspectives Quarterly* 17, no. 3: 12–14. http://dx.doi.org/10.1111/j.1540-5842.2004.00693.x.

Joy, Bill. 2000. "High Technology's Dark Side." Interview, 10 May. http://www.pbs.org/newshour/bb/cyberspace/jan-june00/dark_5-10.html.

Joy, Bill. 2000. "Why the Future Doesn't Need Us: Our Most Powerful 21st Century Technologies – Robotics, Genetic Engineering, and Nanotech – Are Threatening to Make Humans an Endangered Species." *Wired*, April: 238–62.

Jünger, Ernst. 1932. "Technology as the Mobilization of the World through the Gestalt of the Worker." Reprinted in Carl Mitcham and Robert Mackey, eds, *Philosophy and Technology: Readings in the Philosophical Problems of Technology*. New York: Free Press, 1983.

Kant, Immanuel. 1959. *Foundations of the Metaphysics of Morals*, trans. Lewis White Beck. New York: Liberal Arts Press.

Kant, Immanuel. 2002. *Groundwork of the Metaphysics of Morals*. In *Immanuel Kant: Groundwork of the Metaphysic of Morals in Focus*, ed. Lawrence Pasternack, trans. H.J. Patton, 17–98 New York: Routledge.

Kass, Leon. 1971. "The New Biology: What Price Relieving Man's Estate?" *Science* 174, no. 4011: 779–88.

Kass, Leon. 1985. *Toward a More Natural Science: Biology and Human Affairs*. New York: Free Press.

Kass, Leon. Spring 1992. "Organs for Sale? Propriety, Property, and the Price of Progress." *Public Interest* 107: 65–86. Medline:10118413.

Kass, Leon. 1993. "The Problem of Technology." In *Technology in the Western Political Tradition*, ed. Arthur Melzer, Jerry Weinberger, and M. Richard Zinman, 1–24. Ithaca: Cornell University Press.

Kass, Leon. 1997. "Testimony in Front of the National Bioethics Advisory Commission." 14 March, Watergate Hotel, Washington, DC. http://bioethics.georgetown.edu/nbac/transcripts/1997/3-14-97.pdf.

Kass, Leon. 2004. *Life, Liberty and the Defense of Dignity: The Challenge for Bioethics*. San Francisco: Encounter Books.

Kramer, Peter D. 1993. *Listening to Prozac*. New York: Penguin Books.

Kupperman, Joel. 1991. *Character*. Oxford: Oxford University Press.

von Linné, Carl. 1806. *A General System of Nature through the Three Grand Kingdoms of Animals, Vegetables, and Minerals: Classes, orders, genera, species, and varieties, with their habitation, manners, economy, structure and peculiarities*. 13th ed. 7 vols. London: Lackington, Allen, and Co.

Livingston, Edward H. 2007. "Development of Bariatric Surgery-specific Risk Assessment Tool." *Surgery for Obesity and Related Diseases* 3, no. 1: 14–20, discussion 20. http://dx.doi.org/10.1016/j.soard.2006.10.009. Medline:17196436.

Löwith, Karl. 1988. "My Last Meeting with Heidegger in Rome, 1936." *New German Critique*, no. 45, special issue on Bloch and Heidegger, 115–16.

Reprinted in *The Heidegger Controversy: A Critical Reader*, ed. Richard Wolin, 140–3. Cambridge: MIT Press, 1993.

Luther, Martin. 1915. "Ninety-Five Theses." In *Works of Martin Luther*, ed. and trans. Adolph Spaeth et al., vol. 1, 29–38. Philadelphia: A.J. Holman Company.

Machiavelli, Niccolò. 1989. *The Chief Works and Others: Volumes I and II*, trans. Allan H. Gilbert. Durham, NC: Duke University Press.

MacIntyre, Alasdair. 1984. *After Virtue: A Study in Moral Theory*. 2nd ed. South Bend: University of Notre Dame Press.

MacIntyre, Alasdair. 2007. *After Virtue: A Study in Moral Theory*. South Bend: University of Notre Dame Press.

Mahowald, Mary B., and Anthony P. Mahowald. Winter 2002. "Embryonic Stem Cell Retrieval and a Possible Ethical Bypass." *American Journal of Bioethics* 2, no. 1: 42–3. http://dx.doi.org/10.1162/152651602317267871. Medline:12085945.

Marcuse, Herbert. 1993. "An Exchange of Letters." In *The Heidegger Controversy: A Critical Reader*, ed. Richard Wolin, 152–64. Cambridge: MIT Press.

Marcuse, Herbert. 1966. *One-Dimensional Man*. Boston: Beacon Press.

McKinnon, Christine. 1999. *Character, Virtue Theories, and the Vices*. Peterborough, ON: Broadview Press.

McLuhan, Marshall. 1964. *Understanding Media: The Extensions of Man*. New York: McGraw-Hill.

McLuhan, Marshall. 1992. *The Gutenberg Galaxy*. Toronto: University of Toronto Press.

McLuhan, Marshall. 1995. "Playboy Interview." In *Essential McLuhan*, ed. Eric McLuhan and Frank Zingrone, 233–69. Concord, ON: Anansi.

Melzer, Arthur. 1993. "The Problem with the 'Problem of Technology.'" In *Technology and the Western Political Tradition*, ed. Arthur Melzer, Jerry Weinberger, and M. Richard Zinman, 287–321. Ithaca: Cornell University Press.

Mesthene, Emmanuel. 1970. *Technological Change: Its Impact on Man and Society*. New York: Signet.

Mommsen, Theodor, and Paul Krueger, eds. *The Digest of Justinian*. English translation edited by Alan Watson. 1985. 4 vols. Philadelphia: University of Pennsylvania Press.

Mooney, Chris. 2001. "Irrationalist in Chief." *American Prospect* 12, 17. http://prospect.org/article/irrationalist-chief.

Mumford, Lewis. 1974. *Pentagon of Power: The Myth of the Machine*. Vol. 2. New York: Harcourt Brace Jovanovich.

Murdoch, Iris. 1970. *The Sovereignty of Good*. New York: Routledge.

National Center for Health Statistics Health, United States. 2007. *Chartbook on Trends in the Health of Americans*. Hyattsville, MD.

Newberg, Andrew, MD, Eugene G. D'Aquili, and Vince Rause. 2001. *Why God Won't Go Away: Brain Science and the Biology of Belief*. New York: Random House, Inc.

Newberg, Andrew, MD, and Mark Robert Waldman. 2006. *Why We Believe What We Believe: Uncovering Our Biological Need for Meaning, Spirituality, and Truth*. New York: Free Press.

Obama, Barack. 2009. Press conference, 9 February.

Olafson, Frederick A. 2000. "Heidegger's Thought and Nazism." *Inquiry* 43, no. 3: 271–88. http://dx.doi.org/10.1080/002017400414863.

Orwell, George. 1949. *Nineteen Eighty-Four*. New York: Penguin Group (USA), Inc.

Palahniuk, Chuck. 1996. *Fight Club*. New York: Henry Holt and Co., LLC.

Patten, S.B., J.V. Williams, J. Wang, C.E. Adair, R. Brant, A. Casebeer, and C. Barbui. 2005. "Antidepressant Pharmacoepidemiology in a General Population Sample." *Journal of Clinical Psychopharmacology* 25, no. 3: 285–7. http://dx.doi.org/10.1097/01.jcp.0000162815.07442.5a. Medline:15876915.

Plato. 1873. "The Statesman." In *The Dialogues of Plato*. Trans. Benjamin Jowett. New York: Scribner, Armstrong and Co.

Plato. 1993. *Sophist*. Indianapolis: Hackett Publishing Co., Inc.

Plato. 2003. *The Republic*. London: Penguin Books Ltd.

Polenberg, Richard, ed. 2002. *In the Matter of J. Robert Oppenheimer: The Security Clearance Hearing*. Ithaca: Cornell University Press.

The President's Council on Bioethics. 2002. *Human Cloning and Human Dignity: An Ethical Inquiry*. Washington, DC.

Quetelet, Adolphe. 1849. *Popular Instructions on the Calculation of Probabilities*. Trans. R Beamish. London: J. Weale.

Quetelet, Adolphe. 1842. *A Treatise on Man and the Development of His Faculties*. Edinburgh: William and Robert Chambers. Reprinted in "Quetelet on the Study of Man." *Population and Development Review* 22, no. 3 (1996): 547–55.

Rockmore, Tom. 1991. *On Heidegger's Nazism and Philosophy*. Berkeley: University of California Press.

Rockmore, Tom. 1992. *The Heidegger Case: On Philosophy and Politics*. Philadelphia: Temple University Press.

Rosen, Stanley. 1993. "*Techne* and the Origins of Modernity." In *Technology in the Western Political Tradition*, ed. Arthur Melzer, Jerry Weinberger, and M. Richard Zinman, 69–84. Ithaca: Cornell University Press.

Rowley, Janet D., Elizabeth Blackburn, Michael S. Gazzaniga, and Daniel W. Foster. 2002. "Harmful Moratorium on Stem Cell Research." *Science* 297, no. 5589: 1957. http://dx.doi.org/10.1126/science.297.5589.1957. Medline:12242409.

Ruderman, Richard S. 1997. "Aristotle and the Recovery of Political Judgment." *American Political Science Review* 91, no. 2: 409–20. http://dx.doi.org/10.2307/2952364.

Ruskin, John. 1876. "A Protest against the Extension of Railways in the Lake District." In *The Works of John Ruskin*, ed. E.T. Cook and Alexander Wedderburn, vol. 34. London: George Allen.

Sale, Kirkpatrick. 1995. "Lessons from the Luddites: Setting Limits on Technology." *The Nation* 260, no. 22: 785–88.

Schaff, Philip, ed. 1994. "The Lives of the Fathers" (*Vitae Patrum*). In *Nicene and Post-Nicene Fathers*. Grand Rapids, MI: W.M.B. Eerdmans Publishing Co. http://www.ccel.org/ccel/schaff/npnf206.toc.html.

Schirmacher, Wolfgang. 1983. *Technik und Gelassensheit*. Freiburg: Alber.

Schofield, Norman. 2005. "The Intellectual Contribution of Condorcet to the Founding of the US Republic 1785–1800." *Social Choice and Welfare* 25, nos. 2–3: 303–18. http://dx.doi.org/10.1007/s00355-005-0005-y.

Schwartz Cowan, R. 1972. "Francis Galton's Statistical Ideas: The Influence of Eugenics." *Isis* 63, no. 4: 509–28. http://dx.doi.org/10.1086/351000. Medline:4572701.

Shelley, Mary. 1888. *Frankenstein or The Modern Prometheus*. London: George Routledge and Sons.

Shipley, Amy. 2005. "Chemists Stay a Step Ahead of Drug Testers." *Washington Post*, 18 October: E1.

Spengler, Oswald. 1971. *Der Mensch und die Technik: Beitrag zu einer Philosophic des Lebens*. 1931. Munich: C.H. Beck'sche Verlagsbuchhandlung.

Spengler, Oswald. 1991. *The Decline of the West*. Abr. ed. Oxford: Oxford University Press.

"Statement of the Catholic Leadership Conference on Human Cloning." 2001. http://www.priestsforlife.org/articles/01-11-01humancloningclc.htm.

Steinberg, Theodore. 2004. *Nature Incorporated*. Cambridge: Cambridge University Press.

Steinberger, Peter. 1993. *The Concept of Political Judgment*. Chicago: University of Chicago Press.

Steiner, George. 1987. *Martin Heidegger*. Chicago: University of Chicago Press.

Thevis, M., and W. Schänzer. May 2007. "Emerging Drugs – Potential for Misuse in Sport and Doping Control Detection Strategies." *Mini-Reviews in Medicinal Chemistry* 7, no. 5: 533–79. http://dx.doi.org/10.2174/138955707780619590. Medline:17504189.

Thomson, Iain. 2005. *Heidegger on Ontotheology*. Cambridge: Cambridge University Press. http://dx.doi.org/10.1017/CBO9780511499210.

Thoreau, Henry David. 1986. *Walden and Civil Disobedience*. New York: Penguin.

Thucydides. 1972. *History of the Peloponnesian War*. Trans. Rex Warner. London: Penguin.

Towbin, A., T.H. Inge, V.F. Garcia, H.R. Roehrig, R.H. Clements, C.M. Harmon, and S.R. Daniels. 2004. "Beriberi after Gastric Bypass Surgery in Adolescence." *Journal of Pediatrics* 145, no. 2: 263–7. http://dx.doi.org/10.1016/j.jpeds.2004.04.051. Medline:15289782.

Turgot, Anne-Robert. 1987. "Memorandum on Local Government." In *The Old Regime and the French Revolution*, ed. K.M. Baker, 97–117. Chicago: University of Chicago Press.

Vogel, Lawrence. 2006. "Natural Law Judaism? The Genesis of Bioethics in Hans Jonas, Leo Strauss, and Leon Kass." *Hastings Center Report* 36, no. 3: 32–44. http://dx.doi.org/10.1353/hcr.2006.0051. Medline:16776021.

Wachbroit, Robert. 1997. "Genetic Encores: The Ethics of Human Cloning." *Report from the Institute for Philosophy and Public Policy* 17, no. 4: 1–7.

Weil, Eric. 1947. "Le cas Heidegger." *Les Temps Modernes* 22 (July): 128–38.

Weiss, Rick. 2004. "Bush Ejects Two from Bioethics Council: New Criticism That President Puts Politics Ahead of Science." *Washington Post*, 28 February: A06.

Winner, Langdon. 1977. *Autonomous Technology: Technics-Out-of-Control as a Theme in Political Thought*. Cambridge, MA: MIT Press.

Wolin, Richard, ed. 1993. *The Heidegger Controversy: A Critical Reader*. Cambridge: MIT Press.

Wordsworth, William, and Samuel Coleridge. 1798. "The Tables Turned." In *Lyrical Ballads, with a Few Other Poems*, 186–8. London: J. & A. Arch, Gracechurch Street.

Zimmerman, Michael. 1990. *Heidegger's Confrontation with Modernity*. Bloomington: Indiana University Press.

Zuckert, Catherine. 1990. "Martin Heidegger: His Philosophy and His Politics." *Political Theory* 18, no. 1: 51–79. http://dx.doi.org/10.1177/0090591790018001004.

Index

Lightning Source UK Ltd.
Milton Keynes UK
UKHW041437161122
412297UK00015B/56